Geometry of the Quintic

Geometry of the Quintic

Jerry Shurman

With illustrations by Josh Levenberg

A Wiley-Interscience Publication
JOHN WILEY & SONS, INC.
New York • Chichester • Brisbane • Toronto • Singapore • Weinheim

This text is printed on acid-free paper.

Library of Congress Cataloging in Publication Data:

Shurman, Jerry Michael.

Geometry of the quintic / Jerry Shurman ; with illustrations by
Josh Levenberg

p. cm.

"A Wiley-Interscience publication."

Includes bibliographical references and index.

ISBN 0-471-13017-6

1. Quintic equations. 2. Curves, Quintic. I. Title.

QA215.S48 1997

512.9'42—DC20 96-43766

Printed in the United States of America

10 9 8 7 6 5 4 3

Preface

A body of mathematics often comes into sharp focus for the student who is learning it for the second time, to solve an engaging problem. My aim in writing this book is to provide such a review and application for an advanced undergraduate or beginning graduate audience. The material is taken from the classic 1884 text *Lectures on the Icosahedron and Equations of the Fifth Degree* by Felix Klein [Kl], and from the 1989 paper "Solving the quintic by iteration" by Peter Doyle and Curt McMullen [Do-Mc]. The mathematical curriculum has so evolved since Klein's day that his subject—traversing geometry, linear algebra, group theory, complex analysis and Galois theory—now provides a wonderful consolidation of an undergraduate education, and using the icosahedron to solve the quintic supplies a fascinating problem indeed. Unfortunately, even in English translation the century-old syntax of Klein's book is tough going for a contemporary student. (L. E. Dickson's 1926 *Modern Algebraic Theories* [Di] covers some of the same material more accessibly, but it eschews higher mathematics in favor of an elementary presentation.) Add to Klein's work the wrinkle of Doyle and McMullen's recent result, which updates the classical solution of the quintic by transcendental functions to a solution by pure iteration, and this text at this level fills a pedagogical gap and is of mathematical interest.

The prerequisites for this book are semester courses in linear algebra (including inner product spaces), real analysis (including point set topology in Euclidean space), complex analysis (including theory of analytic and meromorphic functions and conformal mappings), and modern algebra (including symmetric groups and acquaintance with rings and fields). Some beginnings of general topology, Galois theory, and (to a lesser extent) algebraic geometry will help one read the book in its intended review-and-apply spirit, but for an ambitious reader these basics are not required since they are presented before being used. In any case, general mathematical maturity can compensate for not having seen a particular topic.

My expository goal in reviewing undergraduate material is to present it "right" for the second-time reader, in the senses of

- showing by concise summary that bodies of material that looked big the first time through aren't so formidable after all,
- providing, via exercises, a workbook environment for going actively through material that was previously read more passively,
- applying reviewed concepts from various areas in the context of one problem, to demonstrate that using general theorems is helpful and easy,
- presenting familiar subjects in nonredundant, forward-looking fashion, by highlighting aspects that may be slighted in a first presentation and by using graduate language.

Since long expositions can camouflage key ideas, the writing here is deliberately terse: this book is therefore meant to be read slowly, in small doses. I make no effort to give exhaustive treatment of every topic that this material touches on, especially topics already well covered in other contemporary texts. The main narrative suitably illuminates the varied terrain it traverses.

Chapter 1 puts structure on the sphere. As the one-point compactification of the complex plane, the Riemann sphere is well suited to complex analysis. And since all meromorphic functions on the Riemann sphere are in fact rational, they are conveniently viewed as algebraic mappings of the complex projective line, which is yet another version of the sphere. Chapter 2 classifies the finite groups of automorphisms (invertible meromorphic self-maps) of the Riemann sphere. Up to conjugacy, these are the cyclic groups, the dihedral groups, and the rotation groups of the Platonic solids. The cyclic and dihedral groups may also be viewed as the rotation groups of certain figures, and conversely the algebraic structure of the Platonic rotation groups is well known: they are the alternating group A_4 for the tetrahedron, the symmetric group S_4 for the octahedron/cube, and the alternating group A_5 for the icosahedron/dodecahedron. Chapter 3 computes for each such group Γ a generator of the field of Γ-invariant rational functions. Chapter 4 discusses algebraic aspects of inverting this generator, a problem equivalent to constructing a certain field extension with Galois group Γ. Since all of the groups catalogued in Chapter 2 except the icosahedral group are solvable, Galois theory shows that radicals (i.e., nth root extractions) invert the generator in all but the icosahedral case. Inverting the icosahedral generator turns out to be equivalent to solving a certain quintic polynomial, the Brioschi quintic, which is not solvable by radicals. Chapter 5 reduces the general quintic to Brioschi form. Carrying out this reduction requires a square root that is auxiliary to the set-up of the general quintic. Chapter 6 proves Kronecker's Theorem, which asserts that the auxiliary square root is necessary for any such reduction. Chapter 7 discusses Doyle and McMullen's solution of the Brioschi quintic by purely iterative algorithm.

While it may be unduly optimistic for me to envision this book as the summer poolside reading rage for budding mathematicians between college and graduate school, I do hope that individual students and student seminars will use it for independent study. To this end, related readings are recommended at the beginning of each chapter. The exercises intertwine closely with the exposition and are cited freely, making them an integral part of the text. They should all be contemplated, even if not solved. Most of the chapters begin with general theorems and then proceed to specific calculations, which are meant to keep the material vividly explicit. These calculations get quite elaborate, so the book may be more profitably (and less laboriously) read with the help of a computer algebra package such as Maple or Mathematica. The sophistication of the material increases through the book, with Chapters 6 and 7 considerably more advanced than the others.

The work here is largely a collaboration with Reed College students, whose contribution is a pleasure to acknowledge. James T. Brown and Bruce Fields wrote their senior theses on portions of the material. Ben Davis, Mark Jefferys, Josh Levenberg, Ye Li, Zeb Strong, and Greg Vande Krol suggested many improvements to the course notes that have evolved into this text. Douglas Squirrel worked carefully through a near-final draft. And Josh Levenberg produced the more attractive computer illustrations using a ray-tracing program. Among the many colleagues who helped me with the project, Joe Buhler, Cris Poor, and especially David Cox merit special thanks for their remarkable generosity.

Contents

CHAPTER 1

The complex sphere

The first step toward studying the symmetries and function theory of the sphere is to supplement its geometric definition with additional structure. This chapter defines the Riemann sphere and the complex projective line, the first suitable for complex analytic methods, the second for algebraic ones, and both equivalent to the sphere. The equivalence is topological: the sphere maps bijectively to each via a homeomorphism, meaning a continuous map with continuous inverse. The Riemann sphere consists of the complex numbers and one more point. It is not geometrically round but it is conformally equivalent to the sphere, meaning that angles are preserved under the topological map between them. The meromorphic functions on the Riemann sphere, superficially analytic objects, are in fact the rational functions in one variable, which are algebraic. These are in turn conveniently represented as pairs of homogeneous polynomials in two variables, describing self-maps of the complex projective line.

Recommended reading: Chapter 1 of Jones and Singerman [Jo-Si] discusses the sphere, overlapping some with the discussion here. Also, the first appendix to [Jo-Si] provides a quick review of complex analysis. For more review, see Ahlfors [Ah], Marsden and Hoffman [Ma-Ho], or any of countless other good complex analysis texts for a development of analytic and meromorphic functions and conformal mappings.

1. Topological preliminaries

Euclidean space $\mathbf{R}^3$ with points $p = (p_1, p_2, p_3)$ and the usual absolute value $|p| = \sqrt{p_1^2 + p_2^2 + p_3^2}$ contains the sphere $S^2 = \{p : |p| = 1\}$ as a subset. This chapter lays the groundwork for a careful study of the sphere by constructing two other objects equivalent to it under continuous mappings. The first order of business is therefore to discuss continuity.

The definition of continuity for mappings between Euclidean spaces relies on the absolute value

$$|p| = \sqrt{\sum_{i=1}^{n} p_i^2} \qquad \text{for } p = (p_1, \ldots, p_n) \in \mathbf{R}^n.$$

Specifically, let $f : \mathbf{R}^n \longrightarrow \mathbf{R}^m$ be a mapping. By definition f is continuous if the following condition holds for each point $p \in \mathbf{R}^n$: given $\varepsilon > 0$ there exists $\delta > 0$ such that for $x \in \mathbf{R}^n$,

$$|x - p| < \delta \quad \Longrightarrow \quad |f(x) - f(p)| < \varepsilon.$$

But not all mathematical spaces are Euclidean, so this definition is limited in scope.

Rephrasing continuity in the language of point set topology generalizes it widely and illuminates its properties. The absolute value allows the definition of open set. For $p \in \mathbf{R}^n$ and $r > 0$, the **ball** of radius r about p is

$$B(p, r) = \{x \in \mathbf{R}^n : |x - p| < r\}.$$

A set $O \subset \mathbf{R}^n$ is **open** if it is a union of open balls, or equivalently, for each point $p \in O$ some ball $B(p, r)$ lies in O. (Exercise 1.1.1 asks for a proof of the equivalence.) These open sets are called the **usual topology on $\mathbf{R}^n$** and have the following properties (Exercise 1.1.2):

1. $\emptyset$ and $\mathbf{R}^n$ are open,
2. any union of open sets is open,
3. any finite intersection of open sets is open.

The key idea is that continuity can be expressed in terms of open sets. Again take $f : \mathbf{R}^n \longrightarrow \mathbf{R}^m$. The **inverse image** of any set $T \subset \mathbf{R}^m$ is the set of points in $\mathbf{R}^n$ that map to it under f,

$$f^{-1}(T) = \{x \in \mathbf{R}^n : f(x) \in T\}.$$

Exercise 1.1.3 now shows that f is continuous exactly when

for any open set T in $\mathbf{R}^m$, the inverse image $f^{-1}(T)$ is open in $\mathbf{R}^n$.

This is the topological definition of continuity.

To discuss continuous mappings from the sphere $S^2 \subset \mathbf{R}^3$, consider more generally a subset A of $\mathbf{R}^n$ and a mapping $f : A \longrightarrow \mathbf{R}^m$. The ε-δ definition for f to be continuous on A is exactly as given above except that the test points p and x must now lie in A. Switching to topological language, the open sets of A are specified as all sets

$$O_A = O \cap A$$

where $O \subset \mathbf{R}^n$ is open. This is the **induced topology** on A. In particular, the open subsets of the sphere S^2 are unions of the curved disks obtained by intersecting it with balls in $\mathbf{R}^3$. The three properties of open sets transfer immediately to A from $\mathbf{R}^n$ via intersection (Exercise 1.1.4). The topological definition of continuity equivalent to ε-δ also remains essentially unchanged (Exercise 1.1.5): f is continuous exactly when

for any open set T in $\mathbf{R}^m$, the inverse image $f^{-1}(T)$ is open in A.

In general, topological terminology generalizes Euclidean space and is the appropriate language for discussing continuous mappings. A **topological space** is a set X and a **topology** $\mathcal{T}$, meaning a collection of subsets of X (the **open** sets), such that

1. $\emptyset$ and X are open,
2. any union of open sets is open,
3. any finite intersection of open sets is open.

Strictly speaking, the space should be denoted by the pair $(X, \mathcal{T})$, but when $\mathcal{T}$ is clearly established we simply call the space X. A mapping of topological spaces $f : X \longrightarrow Y$ is **continuous** if

for any open set T in Y, the inverse image $f^{-1}(T)$ is open in X.

A mapping of topological spaces $f : X \longrightarrow Y$ is a **homeomorphism** if it is a bijection and both f and f^{-1} are continuous.

Exercise 1.1.4 generalizes directly to show that any subset W of X is a topological space in its own right with the induced topology it inherits from X via intersection, $\mathcal{T}_W = \{O \cap W : O \subset X \text{ is open}\}$.

A subset of $\mathbf{R}^n$ is **closed** if its complement is open and **bounded** if it sits in some ball. Subsets of $\mathbf{R}^n$ that are both closed and bounded have the following topological characterization (Exercise 1.1.6):

(1.1.1) HEINE–BOREL THEOREM. *Let K be a subset of $\mathbf{R}^n$. Then K is closed and bounded if and only if K satisfies the following condition as a topological space with the induced topology from $\mathbf{R}^n$:*

> (**HB**) *Any collection of open subsets of K whose union is K (an* **open cover** *of K) contains a subcollection consisting of finitely many subsets whose union is still K (a* **finite subcover***).*

A topological space K satisfying (**HB**) is called **compact**. Compact sets in $\mathbf{R}^n$ are easy to recognize by their nontopological characterization; in particular the sphere S^2 is clearly compact. The topological characterization

(**HB**) of compactness is **intrinsic**, meaning it refers to the topological space K but to no external objects. Compactness is invariant under continuity:

(1.1.2) THE CONTINUOUS IMAGE OF A COMPACT SPACE IS COMPACT. *Let K and Y be topological spaces with K compact, and let $f : K \longrightarrow Y$ be continuous. Then $f(K) = \{f(p) : p \in K\}$ is compact.*

Proving this (Exercise 1.1.9) is straightforward once one establishes some basic facts about the behavior of images and inverse images of sets under mappings. These are the subject of Exercise 1.1.8.

The definition of continuity can be turned around to transfer topological structure from one set to another. Let X be a topological space, Y a set, and $f : X \longrightarrow Y$ a surjection. Define open sets in Y by the rule

$$T \subset Y \text{ is open} \qquad \Longleftrightarrow \qquad f^{-1}(T) \subset X \text{ is open.}$$

This **quotient topology** makes Y a topological space, and if f is bijective the quotient topology on Y makes f a homeomorphism (Exercise 1.1.11). In particular, we will use this device in Sections 1.3 and 1.4 to transfer the topology of the sphere S^2 to corresponding bijective images.

Exercises

1.1.1. Show that a set $O \subset \mathbf{R}^n$ is a union of open balls if and only if for each point $p \in O$ some ball $B(p,r)$ lies in O. (This requires the triangle inequality: $|x - z| \leq |x - y| + |y - z|$ for all $x, y, z \in \mathbf{R}^n$.)

1.1.2. Prove the open set properties for $\mathbf{R}^n$.

1.1.3. Prove that topological continuity is equivalent to ε-δ continuity for mappings between Euclidean spaces.

1.1.4. Prove the open set properties for any subset $A \in \mathbf{R}^n$.

1.1.5. Prove that topological continuity is equivalent to ε-δ continuity for mappings $f : A \longrightarrow \mathbf{R}^m$ where $A \subset \mathbf{R}^n$ receives the induced topology from $\mathbf{R}^n$.

1.1.6. Prove the Heine–Borel theorem. (This is fairly substantial. You might want to review from a real analysis book.)

1.1.7. Prove the Bolzano–Weierstrass theorem: A subset K of $\mathbf{R}^n$ is compact if and only if every sequence in K has a subsequence that converges in K.

1.1.8. Let $f : X \longrightarrow Y$ be a mapping between sets, let S and S_i (as i runs through some index set) be subsets of X, and let T and T_i be subsets of Y. Each of the following pairs is plausibly related:

1. $f(\cup S_i)$ and $\cup f(S_i)$,
2. $f(\cap S_i)$ and $\cap f(S_i)$,
3. $f(S^c)$ and $(f(S))^c$ (S^c is the complement of S),
4. $f^{-1}(f(S))$ and S,
5. $f^{-1}(\cup T_i)$ and $\cup f^{-1}(T_i)$,
6. $f^{-1}(\cap T_i)$ and $\cap f^{-1}(T_i)$,
7. $f^{-1}(T^c)$ and $(f^{-1}(T))^c$,
8. $f(f^{-1}(T))$ and T.

Find the relation in each case, making assumptions about f as needed. In each case, what assumption about f gives equality?

1.1.9. Prove that the continuous image of a compact space is compact.

1.1.10. Prove the Extreme Value Theorem: A continuous function from a compact space K to $\mathbf{R}$ takes maximum and minimum values.

1.1.11. Let X be a topological space, Y a set, and $f : X \longrightarrow Y$ a surjection. Define open sets in Y by the rule

$$T \subset Y \text{ is open} \qquad \Longleftrightarrow \qquad f^{-1}(T) \subset X \text{ is open.}$$

Show that this quotient topology makes Y a topological space. If f is bijective, show that the quotient topology on Y makes f a homeomorphism.

2. Stereographic projection

The equatorial plane $\{p \in \mathbf{R}^3 : p_3 = 0\}$ in Euclidean space $\mathbf{R}^3$ naturally identifies with $\mathbf{R}^2$ via $(p_1, p_2, 0) \leftrightarrow (p_1, p_2)$. Let $\mathbf{n} = (0, 0, 1)$ denote the north pole on the sphere $S^2 = \{p \in \mathbf{R}^3 : |p| = 1\}$. **Stereographic projection**

$$\pi : S^2 \setminus \{\mathbf{n}\} \longrightarrow \mathbf{R}^2$$

is defined by intersecting the line through the north pole $\mathbf{n}$ and the input point p with the plane, i.e., $\pi(p) = \ell(\mathbf{n}, p) \cap \mathbf{R}^2$, where $\ell(\mathbf{n}, p) = \{(1-t)\mathbf{n} + tp : t \in \mathbf{R}\}$ is the line through $\mathbf{n}$ and p. (See Figure 1.2.1.)

The point $(1-t)\mathbf{n} + tp$ has last coordinate $1 - t + tp_3$. This equals 0 for $t = 1/(1-p_3)$, when the other coordinates are $p_j/(1-p_3)$ for $j = 1, 2$. Thus

$$\pi(p_1, p_2, p_3) = \left(\frac{p_1}{1-p_3}, \frac{p_2}{1-p_3} \right). \tag{1.2.1}$$

For the inverse map, take $q = (q_1, q_2, 0)$ in the plane. Any point $p = (1-t)\mathbf{n} + tq$ in $\ell(\mathbf{n}, q)$ satisfies $|p|^2 = (1-t)^2 + t^2|q|^2$. This equals 1 for $t = 2/(|q|^2 + 1)$

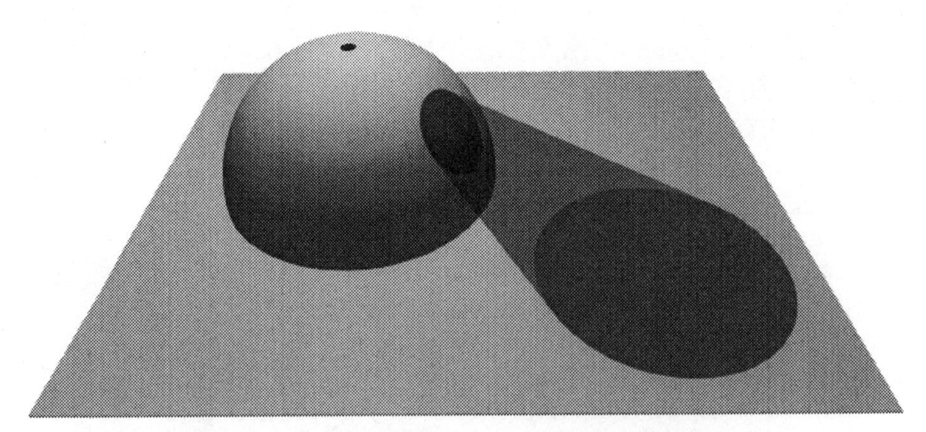

Figure 1.2.1. Stereographic projection

(ignoring $t = 0$, which gives the north pole), showing that

$$\pi^{-1}(q_1, q_2) = \left(\frac{2q_1}{q_1^2 + q_2^2 + 1}, \frac{2q_2}{q_1^2 + q_2^2 + 1}, \frac{q_1^2 + q_2^2 - 1}{q_1^2 + q_2^2 + 1} \right).$$

Both S^2 and $\mathbf{R}^2$ (viewed as the equatorial plane) inherit topologies from their ambient space $\mathbf{R}^3$. The induced topology on $\mathbf{R}^2$ is the usual one by Exercise 1.2.1. Stereographic projection is the restriction of a componentwise rational map from $\{p : p_3 \neq 1\}$ to $\mathbf{R}^2$, so it is continuous by Exercise 1.2.2. The componentwise rational map π^{-1} is also continuous by Exercise 1.2.3, so π is a homeomorphism.

Stereographic projection is conformal, meaning it preserves the angles between curves. To see this, take a point $p \in S^2 \setminus \{\mathbf{n}\}$, let T_p denote the tangent plane to S^2 at p, and let $T_{\mathbf{n}}$ denote the tangent plane to S^2 at $\mathbf{n}$. Working first in the $\mathbf{0n}p$-plane (see Figure 1.2.2), we have equal angles α and right angles between the radii and the tangent planes, hence equal angles β, hence equal angles β', and hence equal lengths b.

Now let γ be a smooth curve on S^2 through p, let t be its tangent at

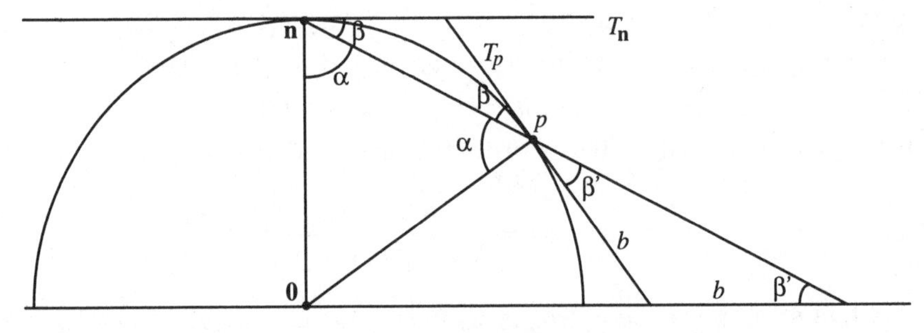

Figure 1.2.2. Side view of stereographic projection

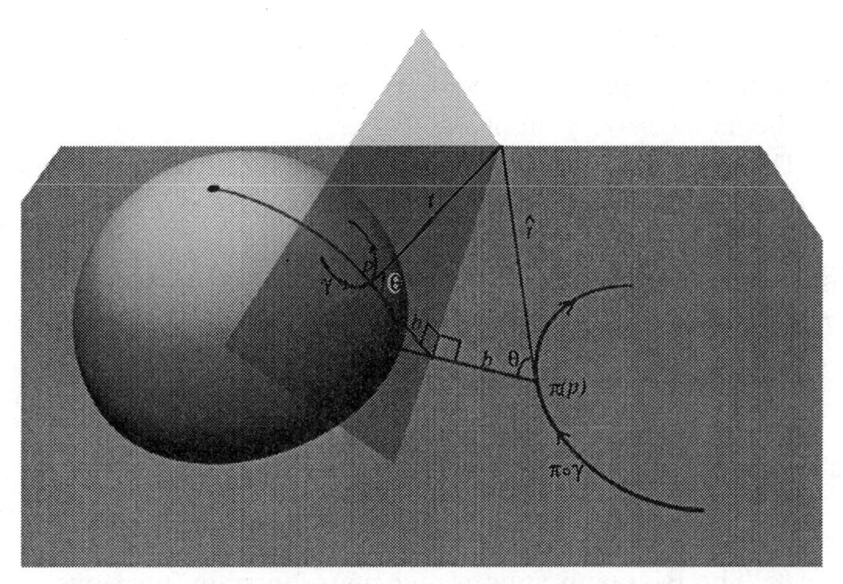

Figure 1.2.3. Stereographic projection is conformal

p, and let $\hat{t}$ be the intersection of the plane containing $\mathbf{n}$ and t with $\mathbf{R}^2$. (See Figure 1.2.3.) In fact $\hat{t}$ is the tangent to $\pi \circ \gamma$ at $\pi(p)$. To see this, recall that π is the restriction of a rational, hence differentiable, map (also called π) from an $\mathbf{R}^3$-neighborhood of p to $\mathbf{R}^2$ that takes t to $\hat{t}$ near p. (A **neighborhood** of a point is an open set containing the point.) Since γ and t are curves in $\mathbf{R}^3$ with the same tangent t at p, $\pi \circ \gamma$ and $\pi \circ t = \hat{t}$ are curves in $\mathbf{R}^2$ with the same tangent at $\pi(p)$. Since $\hat{t}$ is its own tangent at $\pi(p)$, it is also the tangent to $\pi \circ \gamma$ there.

The lengths b are equal, hence so are the angles θ, by right triangles. Repeating this analysis for a second curve $\tilde{\gamma}$ through p completes the proof.

For a continuation of this argument, showing that stereographic projection takes circles to circles, see Hilbert and Cohn-Vossen [Hi-Co].

Exercises

1.2.1. For $1 \le k < n$, $\mathbf{R}^k$ embeds canonically in $\mathbf{R}^n$ under the map $x \mapsto (x, \mathbf{0}_{n-k})$. Show that the embedded copy of $\mathbf{R}^k$ inherits its usual topology as a subspace of $\mathbf{R}^n$.

1.2.2. Let X and Y be topological spaces and W a subspace of X. Show that if $f : X \longrightarrow Y$ is continuous then the restriction $f_W : W \longrightarrow Y$ is also continuous.

1.2.3. Let X and Y be topological spaces and W a subspace of Y. Show that if $f : X \longrightarrow W$ is continuous as a mapping to W if and only if it is continuous as a mapping to Y.

1.2.4. Illustrate the proof that stereographic projection is conformal when p lies in the lower hemisphere.

1.2.5. The proof that stereographic projection is conformal tacitly assumed that t and $\hat{t}$ meet. Must they? What happens to the proof if they don't?

1.2.6. Show that stereographic projection takes circles to circles.

1.2.7. Generalize stereographic projection to a homeomorphism

$$\pi_n : S^n \setminus \{\mathbf{n}\} \xrightarrow{\sim} \mathbf{R}^n,$$

where now $\mathbf{n} = (0, \ldots, 0, 1) \in \mathbf{R}^{n+1}$.

3. The Riemann sphere and meromorphic functions

Identify the complex number field $\mathbf{C}$ with the plane $\mathbf{R}^2$ via $x+iy \leftrightarrow (x, y)$ (thus $\mathbf{C}$ is now a topological space) and let $\widehat{\mathbf{C}}$ denote the set $\mathbf{C} \cup \infty$, where ∞ is a formal point not in $\mathbf{C}$. This is the **Riemann sphere**.

Many complex analysis books develop meromorphic functions on $\widehat{\mathbf{C}}$ as follows: If $D \setminus \{p\}$ is a punctured disk about a point $p \in \mathbf{C}$, and $f : D \setminus \{p\} \longrightarrow \mathbf{C}$ is analytic, then f has a Laurent expansion

$$f(z) = \sum_{n=-\infty}^{\infty} a_{p,n}(z-p)^n \qquad \text{for all } z \in D \setminus \{p\}.$$

When the Laurent series is truncated from the left, i.e., $a_{p,n} = 0$ for all n less than some N, the function f is called meromorphic at p. If the function $g(z) = f(1/z)$ is meromorphic at 0 then f is called meromorphic at ∞. (Thus the only entire functions meromorphic at ∞ are polynomials—this is Exercise 1.3.1.) A function that is meromorphic at each point $p \in \mathbf{C}$ and at ∞ is called meromorphic on $\widehat{\mathbf{C}}$. Note that as things stand, a meromorphic function on $\widehat{\mathbf{C}}$ need not even be defined on all of $\widehat{\mathbf{C}}$, though we can use the Laurent expansion to patch in values—possibly including ∞—where it isn't. The value ∞ interacts algebraically with $\mathbf{C}$ by rules such as $\infty + z = \infty$ for all $z \in \mathbf{C}$, $\infty \cdot z = \infty$ for all $z \in \mathbf{C} \setminus \{0\}$, etc.

This all feels a bit ad hoc. Giving the Riemann sphere a topology naturally via stereographic projection improves matters by allowing an intrinsic definition of meromorphy that generalizes to other situations. So, extend stereographic projection to a bijection $\pi : S^2 \longrightarrow \widehat{\mathbf{C}}$ by defining, according to (1.2.1) and the identification of $\mathbf{R}^2$ with $\mathbf{C}$,

$$\pi(p) = \begin{cases} \dfrac{p_1 + ip_2}{1 - p_3} & \text{if } p \in S^2 \setminus \{\mathbf{n}\}, \\ \infty & \text{if } p = \mathbf{n}. \end{cases}$$

Give $\widehat{\mathbf{C}}$ the quotient topology by defining

$$T \subset \widehat{\mathbf{C}} \text{ is open} \qquad \Longleftrightarrow \qquad \pi^{-1}(T) \subset S^2 \text{ is open.}$$

The Riemann sphere is now a topological space and extended stereographic projection is innately a homeomorphism.

Being homeomorphic to S^2, $\widehat{\mathbf{C}}$ is topologically indistinguishable from a sphere (justifying its name) and compact. Also, $\widehat{\mathbf{C}}$ contains $\mathbf{C}$, with its usual topology, as a subspace. The neighborhoods of ∞ in $\widehat{\mathbf{C}}$ are $\widehat{\mathbf{C}} \setminus K$ where K is a compact subset of $\mathbf{C}$, since the neighborhoods of $\mathbf{n}$ in S^2 are $S^2 \setminus K$ where K is a compact subset of $S^2 \setminus \{\mathbf{n}\}$. The inversion map $z \mapsto 1/z$ on $\mathbf{C} \setminus \{0\}$ corresponds to 180-degree rotation r of S^2 about the x_1-axis (showing this is Exercise 1.3.2), so this map extends continuously to $\widehat{\mathbf{C}}$ by $1/0 = \infty$ and $1/\infty = 0$ since the rotation of S^2 exchanges the poles $\mathbf{n} = \pi^{-1}(\infty)$ and $\mathbf{s} = \pi^{-1}(0)$.

The Riemann sphere $\widehat{\mathbf{C}}$ looks like $\mathbf{C}$ in the small; it is a **manifold** or **Riemann surface**. To quantify this idea, define for each point $p \in \widehat{\mathbf{C}}$ the **local coordinate function**

$$\begin{aligned} c_p &: \widehat{\mathbf{C}} \setminus \{\infty\} \longrightarrow \mathbf{C} \quad \text{by } c_p(z) = z - p \quad \text{if } p \in \mathbf{C}, \\ c_\infty &: \widehat{\mathbf{C}} \setminus \{0\} \longrightarrow \mathbf{C} \quad \text{by } c_\infty(z) = 1/z \quad \text{if } p = \infty. \end{aligned}$$

Each c_p is a homeomorphism that takes p to 0. These coordinates let us use the convenient language of $\mathbf{C}$ to phrase local notions about $\widehat{\mathbf{C}}$. For example, extended stereographic projection π is also conformal at its new domain point $\mathbf{n}$, provided we interpret the statement to apply to $c_\infty \circ \pi$. This follows from the relation $c_\infty \circ \pi = \pi \circ r$, where r is the 180-degree rotation of S^2 mentioned above (which is certainly conformal), since π is conformal at the south pole $r(\mathbf{n})$. (See Exercise 1.3.2.)

Using this new language, define a function $f : \widehat{\mathbf{C}} \longrightarrow \widehat{\mathbf{C}}$ to be **meromorphic** if $f = 0$ (the zero function), or f is locally a nonzero analytic function times an integer power of the local variable. The second case means that for each $p \in \widehat{\mathbf{C}}$ there exist a neighborhood U_p, an integer N, and an analytic function $\phi_p : c_p(U_p) \longrightarrow \mathbf{C}$ with $\phi_p(0) \neq 0$ such that

$$f(z) = c_p(z)^N \phi_p(c_p(z)) \qquad \text{for all } z \in U_p. \tag{1.3.1}$$

Here $0^N\phi_p(0)$ is defined as ∞ if $N < 0$. Since analytic functions are represented by power series, this definition recovers that every nonzero meromorphic f has a suitable Laurent expansion in $c_p(z)$ for z near each p, meaning there exist a neighborhood U_p, an integer N, and complex coefficients $\{a_{p,N}, a_{p,N+1}, a_{p,N+2}, \ldots\}$ with $a_{p,N} \neq 0$ such that

$$f(z) = \sum_{n=N}^{\infty} a_{p,n}c_p(z)^n \qquad \text{for all } z \in U_p.$$

Thus the convention is that $\sum_{n=N}^{\infty} a_{p,n}0^n = \infty$ if $N < 0$. The integer N is called the **order** of f at p, written $\mathrm{ord}_p(f)$. The zeros of f are at the points p where $\mathrm{ord}_p(f) > 0$, and the poles (points mapping to ∞) are at p where $\mathrm{ord}_p(f) < 0$. Since the zeros and poles are isolated (this follows from the definition of nonzero meromorphic function) and $\widehat{\mathbf{C}}$ is compact, there are finitely many of each, by the Bolzano–Weierstrass theorem (Exercise 1.1.7).

The meromorphic functions on $\widehat{\mathbf{C}}$ form a field $K(\widehat{\mathbf{C}})$ containing a copy of $\mathbf{C}$ as a subfield. Showing this is Exercise 1.3.3.

Every meromorphic function f is **analytic in local coordinates**, meaning that each $p \in \widehat{\mathbf{C}}$ has a neighborhood U_p such that $c_{f(p)} \circ f \circ c_p^{-1} : c_p(U_p) \longrightarrow \mathbf{C}$ is analytic at 0. This is clear if $f = 0$. Otherwise, if f is meromorphic at p then by (1.3.1), $f \circ c_p^{-1} = id^N \cdot \phi_p$ on $c_p(U_p)$ (id is the identity function) for suitable $N \in \mathbf{Z}$, ϕ_p and U_p. Since $f(p) = \infty$ exactly when $N < 0$, it follows that on $c_p(U_p)$,

$$c_{f(p)} \circ f \circ c_p^{-1} = \begin{cases} id^N \cdot \phi_p - f(p) & \text{if } N \geq 0, \\ 1/(id^N \cdot \phi_p) & \text{if } N < 0. \end{cases}$$

This is analytic at 0 in either case, showing that f is locally analytic at p as claimed.

The converse is not quite true. All but one function $f : \widehat{\mathbf{C}} \longrightarrow \widehat{\mathbf{C}}$ that are analytic in local coordinates are meromorphic. Showing this is Exercise 1.3.4. The exceptional function, $f(p) = \infty$ for all $p \in \widehat{\mathbf{C}}$, prevents the locally analytic functions from forming a field.

Since analytic functions are continuous and local coordinate maps are homeomorphisms, it follows that every meromorphic function, being analytic in local coordinates, is continuous.

A function $f : \widehat{\mathbf{C}} \longrightarrow \widehat{\mathbf{C}}$ is called **conformal** if it is conformal in local coordinates, meaning for each $p \in \widehat{\mathbf{C}}$ there exists a neighborhood U_p such that $c_{f(p)} \circ f \circ c_p^{-1} : c_p(U_p) \longrightarrow \mathbf{C}$ is conformal at 0. Recall from complex analysis that a function is conformal at a point if and only if it is analytic with nonzero derivative there. Thus by the preceding discussion, any conformal function f on $\widehat{\mathbf{C}}$ is meromorphic. Since stereographic projection π is conformal, any function f on $\widehat{\mathbf{C}}$ corresponding under π to a conformal motion of the sphere S^2 is meromorphic. In particular, any function on $\widehat{\mathbf{C}}$ induced by a rotation of S^2 is meromorphic. We will discuss these rotations in Chapter 2.

Any rational expression $f(Z) = g(Z)/h(Z)$ (where g and h are polynomials with complex coefficients and no common factor, h is not the zero polynomial, and the upper-case Z is an algebraic symbol) defines a meromorphic function, also called f, on $\widehat{\mathbf{C}}$. To specify f when g is not identically zero, note that for any $p \in \mathbf{C}$, g takes the form $g(Z) = (Z-p)^d\tilde{g}(Z-p)$ where $d \in \mathbf{N}$ and $\tilde{g}$ is a polynomial satisfying $\tilde{g}(0) \neq 0$ (this is Exercise 1.3.5), and similarly $h(Z) = (Z-p)^e\tilde{h}(Z-p)$. So setting

$$\begin{aligned} f(z) = g(z)/h(z) &= (z-p)^{d-e}\tilde{g}(z-p)/\tilde{h}(z-p) \\ &= c_p(z)^{d-e}\phi_p(c_p(z)) \quad \text{for } z \text{ near } p, \end{aligned}$$

where $\phi_p = \tilde{g}/\tilde{h}$, defines a meromorphic function about p. To define f about ∞, write $g(Z) = (1/Z)^d\tilde{g}(1/Z)$ where $d \in -\mathbf{N}$ and again $\tilde{g}$ is a polynomial satisfying $\tilde{g}(0) \neq 0$ (Exercise 1.3.5), and similarly for h, and again setting

$$f(z) = g(z)/h(z) = c_\infty(z)^{d-e}\phi_\infty(c_\infty(z)) \quad \text{for } z \text{ near } \infty$$

does the trick. This is essentially the procedure outlined at the beginning of the section. The actual formula for evaluating f is

$$f(p) = \begin{cases} g(p)/h(p) & \text{if } p \in \mathbf{C} \text{ and } h(p) \neq 0, \\ \infty & \text{if } p \in \mathbf{C} \text{ and } h(p) = 0, \\ 0 & \text{if } p = \infty \text{ and } \deg(g) < \deg(h), \\ \infty & \text{if } p = \infty \text{ and } \deg(g) > \deg(h), \\ a_d/b_d & \text{if } p = \infty \text{ and } \deg(g) = \deg(h) = d, \end{cases}$$

where a_d and b_d are the leading coefficients of g and h; but in practice one writes less pedantically $f(p) = g(p)/h(p)$ with all the cases being tacit.

Let $\mathbf{C}(Z)$ denote the set of formal rational expressions $f(Z) = g(Z)/h(Z)$ as above. This set forms a field.

(1.3.2) THEOREM. *The meromorphic functions on* $\widehat{\mathbf{C}}$ *are precisely the rational functions. That is,* $K(\widehat{\mathbf{C}}) = \mathbf{C}(Z)$.

Note that this theorem identifies a field of functions with a field of formal expressions. From now on we will move freely between these two objects, viewing meromorphic functions as mappings for purposes of analysis and as rational expressions for purposes of algebra.

PROOF. Let f be a given meromorphic function on $\widehat{\mathbf{C}}$ and define a polynomial $h(Z) = \prod_b (Z-b)^{-\mathrm{ord}_b(f)}$, where the product is over poles of f in $\mathbf{C}$. Consider the meromorphic function $g(p) = f(p)h(p)$. This has no poles in $\mathbf{C}$, hence its restriction to $\mathbf{C}$ is entire and has a power series representation, $g(z) = \sum_{n=0}^{\infty} a_n z^n$ for all $z \in \mathbf{C}$. Thus $g(z) = \sum_{n=0}^{\infty} a_n c_\infty(z)^{-n}$ for all $z \in \mathbf{C}$. Since g is meromorphic at ∞, the sum must contain only finitely many terms. So g is a polynomial, and $f = g/h$ is rational. □

(1.3.3) COROLLARY. *For every nonzero meromorphic function f on $\widehat{\mathbf{C}}$,*

$$\sum_{p \in \widehat{\mathbf{C}}} \mathrm{ord}_p(f) = 0.$$

PROOF. Let $f(p) = k\prod_a (p-a)^{e_a} / \prod_b (p-b)^{e_b}$, where a runs through the zeros of f in $\mathbf{C}$ and b runs through the poles of f in $\mathbf{C}$. (Note how the Fundamental Theorem of Algebra is being used here.) Then f has order e_a at each a, $-e_b$ at each b, and zero at all other $p \in \mathbf{C}$, so $\sum_{p\in\mathbf{C}} \mathrm{ord}_p(f) = \sum_a e_a - \sum_b e_b$. As for $\mathrm{ord}_\infty(f)$, note that

$$f(p) = k \cdot \frac{p^{(\sum_a e_a)} \prod_a (1-a/p)}{p^{(\sum_b e_b)} \prod_b (1-b/p)} = k \cdot c_\infty(p)^{(\sum_b e_b - \sum_a e_a)} \phi_\infty(c_\infty(p))$$

with $\phi_\infty(0) = 1$ (Exercise 1.3.7 asks for the function ϕ_∞). This formula shows that $\mathrm{ord}_\infty(f) = \sum_b e_b - \sum_a e_a$, canceling the other terms. □

Exercises

1.3.1. Use the first definition of meromorphic in the section to show that the only entire functions meromorphic at ∞ are polynomials.

1.3.2. Let $r : S^2 \longrightarrow S^2$ be the 180-degree rotation given by $r(x_1, x_2, x_3) = (x_1, -x_2, -x_3)$. For any $p \in S^2 \setminus \{\mathbf{n}, \mathbf{s}\}$ ($\mathbf{s}$ is the south pole), show that $1/\pi(p) = \pi(r(p))$. In other words, the following diagram commutes, meaning that going either way around it produces the same effect.

$$\begin{array}{ccc} S^2 \setminus \{\mathbf{n},\mathbf{s}\} & \xrightarrow{\ r\ } & S^2 \setminus \{\mathbf{n},\mathbf{s}\} \\ \pi\big\downarrow & & \big\downarrow\pi \\ \mathbf{C}\setminus\{0\} & \xrightarrow{\ 1/z\ } & \mathbf{C}\setminus\{0\} \end{array}$$

What functions on $\mathbf{C} \setminus \{0\}$ correspond to the other two 180-degree rotations of S^2 about coordinate axes?

1.3.3. Prove that the meromorphic functions on $\widehat{\mathbf{C}}$ form a field containing a copy of $\mathbf{C}$ as a subfield. (If you want to cite Theorem 1.3.2, first prove that the product of meromorphic functions is meromorphic, a fact used in its proof.)

1.3.4. Prove that all but one function $f : \widehat{\mathbf{C}} \longrightarrow \widehat{\mathbf{C}}$ that are analytic in local coordinates are meromorphic, the exception being $f(p) = \infty$ for all $p \in \widehat{\mathbf{C}}$.

1.3.5. Let $g(Z)$ be a nonzero polynomial with complex coefficients. Show that for any $p \in \mathbf{C}$, $g(Z) = (Z - p)^d \tilde{g}(Z - p)$ where $d \in \mathbf{N}$ and $\tilde{g}$ is a polynomial satisfying $\tilde{g}(0) \neq 0$. Show that $g(Z) = (1/Z)^d \tilde{g}(1/Z)$ where $d \in -\mathbf{N}$ and again $\tilde{g}$ is a polynomial satisfying $\tilde{g}(0) \neq 0$.

1.3.6. Prove that a meromorphic function f on $\widehat{\mathbf{C}}$ with no poles is constant.

1.3.7. What is the function ϕ_∞ in the proof of Corollary 1.3.3? Why is ϕ_∞ analytic about 0?

1.3.8. Prove that a nonconstant meromorphic function f on $\widehat{\mathbf{C}}$ takes each value in $\widehat{\mathbf{C}}$ the same number of times, counting multiplicity. (Why does it suffice to show that f takes the values 0 and ∞ the same number of times?) What is this number in terms of g, h where $f = g/h$?

4. The complex projective line and algebraic mappings

The **complex projective line** $\mathbf{P}^1(\mathbf{C})$ is topologically equivalent to the Riemann sphere and algebraically convenient. To define it, put an equivalence relation on the set $\mathbf{C}^2$ by identifying all nonzero scalar multiples of each vector:

$$(z_1, z_2) \sim (w_1, w_2) \quad \text{if} \quad (z_1, z_2) = c \cdot (w_1, w_2) \text{ for some } c \in \mathbf{C}^*.$$

Then $\mathbf{P}^1(\mathbf{C})$ is the set of equivalence classes in $\mathbf{C}^2 \setminus \{\mathbf{0}\}$, the class of (z_1, z_2) being denoted $[z_1 : z_2]$. Specifically, $\mathbf{P}^1(\mathbf{C}) = \{[z : 1] : z \in \mathbf{C}\} \cup [1 : 0]$, so as a set $\mathbf{P}^1(\mathbf{C})$ contains a copy of $\mathbf{C}$ and one more point. The natural map $P : \mathbf{C}^2 \setminus \{\mathbf{0}\} \longrightarrow \mathbf{P}^1(\mathbf{C})$ given by $P(z_1, z_2) = [z_1 : z_2]$ gives $\mathbf{P}^1(\mathbf{C})$ the quotient topology,

$$T \subset \mathbf{P}^1(\mathbf{C}) \text{ is open} \quad \Longleftrightarrow \quad P^{-1}(T) \subset \mathbf{C}^2 \setminus \{\mathbf{0}\} \text{ is open.}$$

(Of course, $\mathbf{C}^2$ gets its topology from $\mathbf{R}^4$.) In particular, if S is a subset of $\mathbf{C}$ then $[S : 1]$ is open if and only if $P^{-1}([S : 1])$ is open in $\mathbf{C}^2 \setminus \{\mathbf{0}\}$, and this is equivalent to S being open in $\mathbf{C}$ (Exercise 1.4.1). Thus the topology of $\mathbf{P}^1(\mathbf{C})$ extends that of $\mathbf{C}$, and the coordinate map (akin to c_0 on the

Riemann sphere)

$$c_{[0:1]} : \mathbf{P}^1(\mathbf{C}) \setminus \{[1:0]\} \longrightarrow \mathbf{C} \qquad \text{where } c_{[0:1]}[z:1] = z$$

is a homeomorphism. The construction of $\mathbf{P}^1(\mathbf{C})$ is symmetric in the coordinates z_1 and z_2, so also $\mathbf{P}^1(\mathbf{C}) = \{[1:z] : z \in \mathbf{C}\} \cup [0:1]$, and the coordinate map (akin to c_∞)

$$c_{[1:0]} : \mathbf{P}^1(\mathbf{C}) \setminus \{[0:1]\} \longrightarrow \mathbf{C} \qquad \text{where } c_{[1:0]}[1:z] = z$$

is a homeomorphism.

The bijection $\varphi : \widehat{\mathbf{C}} \longrightarrow \mathbf{P}^1(\mathbf{C})$ given by $\varphi(z) = [z:1]$ for $z \in \mathbf{C}$, $\varphi(\infty) = [1:0]$ is a homeomorphism. To show this, compute that $c_{[0:1]} \circ \varphi$ and $c_{[1:0]} \circ \varphi \circ c_\infty^{-1}$ are both the identity map on $\mathbf{C}$, which is certainly homeomorphic. The various coordinate functions are homeomorphic and $\widehat{\mathbf{C}} = \mathbf{C} \cup c_\infty^{-1}\mathbf{C}$, so the proof is complete by the topological fact shown in Exercise 1.4.2.

Let $\mathbf{C}[Z_1, Z_2]$ denote the ring of polynomials in the unknowns Z_1, Z_2 over $\mathbf{C}$. (As with the rational functions in the preceding section, the upper-case unknowns should be thought of as purely algebraic variables, while lower-case symbols will take honest values.) A polynomial $G \in \mathbf{C}[Z_1, Z_2]$ is **homogeneous** if all of its terms have the same degree, i.e., for some $d \in \mathbf{N}$, $G(Z_1, Z_2) = \sum_{i=0}^{d} a_i Z_1^i Z_2^{d-i}$. Equivalently (Exercise 1.4.3), $G(CZ_1, CZ_2) = C^d G(Z_1, Z_2)$. Homogeneous polynomials are called **forms**. The set of all degree-d forms in Z_1, Z_2 is written $\mathbf{C}_d[Z_1 : Z_2]$ and the set of all forms of any degree is written $\mathbf{C}[Z_1 : Z_2]$. Since the polynomial 0 is a form of every degree, each $\mathbf{C}_d[Z_1 : Z_2]$ is an abelian group. Forms will often be denoted by upper-case characters to distinguish them from polynomials of one variable.

Any pair $(G(Z_1, Z_2), H(Z_1, Z_2))$ of same-degree forms with no common zeros gives rise to an **algebraic mapping**

$$f : \mathbf{P}^1(\mathbf{C}) \longrightarrow \mathbf{P}^1(\mathbf{C}) \qquad \text{by } f[z_1 : z_2] = [G(z_1, z_2) : H(z_1, z_2)].$$

This is well-defined, meaning $[G(z_1, z_2) : H(z_1, z_2)]$ indeed depends only on the equivalence class $[z_1 : z_2]$ (Exercise 1.4.4). Algebraic mappings with H not identically zero correspond naturally to rational (i.e., meromorphic) functions on $\widehat{\mathbf{C}}$ by

$$[G(Z_1, Z_2) : H(Z_1, Z_2)] \mapsto G(Z, 1)/H(Z, 1) \overset{\text{call}}{=} g(Z)/h(Z)$$

and

$$g(Z)/h(Z) \mapsto [Z_2^d g(Z_1/Z_2) : Z_2^d h(Z_1/Z_2)] \overset{\text{call}}{=} [G(Z_1, Z_2) : H(Z_1, Z_2)],$$

where $d = \max\{\deg(g), \deg(h)\}$. To examine the correspondence, it helps to introduce homogenization and dehomogenization operators. For any form $G(Z_1, Z_2)$, define a polynomial $G_*(Z) = G(Z, 1)$. For any polynomial $g(Z)$ define a form $g^*(Z_1, Z_2) = Z_2^{\deg(g)} g(Z_1/Z_2)$. The idea is that the dehomogenization operator throws away Z_2, for example $(Z_1^4 + 12Z_1Z_2^3)_* = Z^4 + 12Z$, while on the other hand the homogenization operator contributes powers of Z_2 to even out the total degrees of terms, for example $(Z^4 + 12Z)^* = Z_1^4 + 12Z_1Z_2^3$. The operations aren't quite inverse to each other, for instance $((Z_1^2Z_2^2 + 12Z_1Z_2^3)_*)^* = (Z^2 + 12Z)^* = Z_1^2 + 12Z_1Z_2$ loses the Z_2^2 that divides all terms of the original form. For all forms G, $\tilde{G}$ and polynomials g, $\tilde{g}$ the following hold: $(G\tilde{G})_* = G_*\tilde{G}_*$, $(g\tilde{g})^* = g^*\tilde{g}^*$, $(g^*)_* = g$, and $(G_*)^* = Z_2^{-e}G$ where e is the highest power of Z_2 dividing G. These are Exercise 1.4.6.

In the language of these operators, the correspondence between algebraic mappings and rational functions becomes

$$[G : H] \mapsto G_*/H_* \tag{1.4.1}$$

and

$$g/h \mapsto \begin{cases} [g^* : Z_2^{\deg(g)-\deg(h)} h^*] & \text{if } \deg(g) \geq \deg(h), \\ [Z_2^{\deg(h)-\deg(g)} g^* : h^*] & \text{if } \deg(h) > \deg(g). \end{cases} \tag{1.4.2}$$

These are inverse to one another (Exercise 1.4.7). Indeed, taking functions to forms to functions does nothing as one simply multiplies by powers of Z_2 and then removes them; taking forms to functions to forms is a little trickier, relying on the fact that Z_2 can only divide one of G and H. The correspondence is natural in that it commutes with the homeomorphism $\varphi : \widehat{\mathbf{C}} \longrightarrow \mathbf{P}^1(\mathbf{C})$ defined earlier, meaning that for any algebraic map $[G : H]$, the following diagram commutes.

$$\begin{array}{ccc} \widehat{\mathbf{C}} & \xrightarrow{G_*/H_*} & \widehat{\mathbf{C}} \\ \varphi \downarrow & & \downarrow \varphi \\ \mathbf{P}^1(\mathbf{C}) & \xrightarrow{[G:H]} & \mathbf{P}^1(\mathbf{C}) \end{array} \tag{1.4.3}$$

Proving this (Exercise 1.4.8) necessarily involves inspecting cases, but once this is done algebraic maps are a more convenient substitute for rational functions because their defining formula doesn't require lots of cases to take ∞ into account.

Exercises

1.4.1. Let S be a subset of $\mathbf{C}$. Show that S is open in $\mathbf{C}$ if and only if $P^{-1}([S:1])$ is open in $\mathbf{C}^2 \setminus \{\mathbf{0}\}$. (This exercise is rather technical. For the "$\Longrightarrow$" direction, suppose $S \subset \mathbf{C}$ is open and let $(z,w) \in P^{-1}([S:1])$. The goal is to find some positive R such that $B((z,w),R) \subset P^{-1}([S:1])$. Since z/w lies in S (why is w nonzero?), some ball $B(z/w,r)$ sits in S. Find an expression in z, w and r that works as R.)

1.4.2. Let X and Y be topological spaces, and $f : X \longrightarrow Y$ a bijection of sets. Suppose $X = \cup O_i$ with each O_i open in X, each $f(O_i)$ open in Y, and each restriction $f_i : O_i \longrightarrow f(O_i)$ a homeomorphism. Show that f is a homeomorphism.

1.4.3. Show that any polynomial $G \in \mathbf{C}[Z_1, Z_2]$ is homogeneous of degree d if and only if $G(CZ_1, CZ_2) = C^d G(Z_1, Z_2)$.

1.4.4. Verify that algebraic mappings as described in the text are well-defined.

1.4.5. Prove Euler's identity: For $G \in \mathbf{C}_d[Z_1 : Z_2]$, $\sum_{i=1}^{2} Z_i D_i G = dG$, where D_i is the ith partial derivative.

1.4.6. Verify that for all forms G, $\tilde{G}$ and polynomials g, $\tilde{g}$ the following hold: $(G\tilde{G})_* = G_*\tilde{G}_*$, $(g\tilde{g})^* = g^*\tilde{g}^*$, $(g^*)_* = g$, and $(G_*)^* = Z_2^{-e}G$ where e is the highest power of Z_2 dividing G.

1.4.7. Show that the correspondences (1.4.1) and (1.4.2) are inverse to one another.

1.4.8. Show that diagram (1.4.3) commutes.

5. Summary

Each of three homeomorphic copies of the sphere—S^2 itself, the Riemann sphere $\widehat{\mathbf{C}}$, and the complex projective line $\mathbf{P}^1(\mathbf{C})$—has its advantages: S^2 is best for thinking geometrically, $\widehat{\mathbf{C}}$ has the same angles between curves as S^2 and is suited for analysis, and the algebraic maps on $\mathbf{P}^1(\mathbf{C})$ are convenient for studying functions without worrying about infinity or cases. From now on we will move freely among the three models of the sphere as convenient.

CHAPTER 2

Finite automorphism groups of the sphere

This chapter shows that the finite groups of meromorphic bijections from the sphere to itself are classified up to conjugacy by rotation groups. The conjugacy classes of rotation groups are restricted to only a few types: cyclic, dihedral, and the rotations of the Platonic solids.

Recommended reading: Chapter 2 of Jones and Singerman [Jo-Si] complements this chapter, containing details and topics not included here. Coxeter's books [Co 1], [Co 2] are lovely sources for pursuing geometry and symmetry.

1. Automorphisms

An **automorphism** of the Riemann sphere is a bijective meromorphic function on $\widehat{\mathbf{C}}$. The set of such automorphisms is denoted $\mathrm{Aut}(\widehat{\mathbf{C}})$. Since any nonconstant meromorphic function $f = g/h$ takes each value in $\widehat{\mathbf{C}}$ the same number of times counting multiplicity, that number being $\max\{\deg(g), \deg(h)\}$ (this was Exercise 1.3.8), $\mathrm{Aut}(\widehat{\mathbf{C}})$ consists of the fractional linear transformations

$$f(z) = \frac{az+b}{cz+d} \qquad \text{with } a, b, c, d \in \mathbf{C} \text{ and } ad - bc \neq 0$$

(the nonzero determinant excludes the constant functions).

The **general linear group** of 2-by-2 complex matrices with nonzero determinant,

$$\mathrm{GL}_2(\mathbf{C}) = \left\{ \begin{bmatrix} a & b \\ c & d \end{bmatrix} : a, b, c, d \in \mathbf{C},\ ad - bc \neq 0 \right\},$$

maps naturally to $\mathrm{Aut}(\widehat{\mathbf{C}})$ by $m \mapsto f_m$, where if $m = \begin{bmatrix} a & b \\ c & d \end{bmatrix}$ then correspondingly $f_m(z) = \dfrac{az+b}{cz+d}$. A calculation (Exercise 2.1.1) shows that for any $m, n \in \mathrm{GL}_2(\mathbf{C})$, $f_{mn} = f_m \circ f_n$, meaning that the image of $\mathrm{GL}_2(\mathbf{C})$ is a group. Any automorphism takes the form f_m for some $m \in \mathrm{GL}_2(\mathbf{C})$, so in fact the image of $\mathrm{GL}_2(\mathbf{C})$ is all of $\mathrm{Aut}(\widehat{\mathbf{C}})$, and the map $\mathrm{GL}_2(\mathbf{C}) \longrightarrow \mathrm{Aut}(\widehat{\mathbf{C}})$ is a surjective group homomorphism. Its kernel is $\mathbf{C}^*I = \{\lambda I : \lambda \in \mathbf{C}^*\}$, where I is the identity matrix (this is Exercise 2.1.2), so by the First Isomorphism Theorem of group theory (Exercise 2.1.3),

$$\mathrm{Aut}(\widehat{\mathbf{C}}) \cong \mathrm{GL}_2(\mathbf{C})/\mathbf{C}^*I \overset{\text{call}}{=} \mathrm{PGL}_2(\mathbf{C}).$$

This is the **projective** general linear group, meaning (as with the complex projective line) equivalence classes modulo scalar multiplication. Thus the elements of $\mathrm{PGL}_2(\mathbf{C})$ are cosets of matrices,

$$\overline{\begin{bmatrix} a & b \\ c & d \end{bmatrix}} = \left\{\lambda \begin{bmatrix} a & b \\ c & d \end{bmatrix} : \lambda \in \mathbf{C}^*\right\},$$

but one usually denotes them with individual matrix representatives. This is all perfectly intuitive: a matrix specifies an automorphism, while an automorphism only specifies a matrix up to scalar multiples since $(\lambda az + \lambda b)/(\lambda cz + \lambda d) = (az+b)/(cz+d)$ for all $\lambda \in \mathbf{C}^*$. The isomorphism $\mathrm{PGL}_2(\mathbf{C}) \xrightarrow{\sim} \mathrm{Aut}(\widehat{\mathbf{C}})$ is handy because multiplying matrices is more convenient than composing maps, but when manipulating matrices in lieu of automorphisms we need to remember to identify complex scalar multiples, which isn't always easy to do by quick inspection.

Normalizing the automorphism-representing matrices to have determinant 1 cuts down the difficulty of recognizing when two of them represent the same map. To do this, introduce the **special** linear group of 2-by-2 complex matrices with determinant 1,

$$\mathrm{SL}_2(\mathbf{C}) = \left\{\begin{bmatrix} a & b \\ c & d \end{bmatrix} : a,b,c,d \in \mathbf{C},\ ad - bc = 1\right\}.$$

Let $G = \mathrm{GL}_2(\mathbf{C})$, $H = \mathrm{SL}_2(\mathbf{C})$, and $K = \mathbf{C}^*I$. Then $K \lhd G$, $G = HK$ and $H \cap K = \{\pm I\}$ (Exercise 2.1.4), so by the Second Isomorphism Theorem (Exercise 2.1.5),

$$\mathrm{PGL}_2(\mathbf{C}) = HK/K \cong H/(H\cap K) = \mathrm{SL}_2(\mathbf{C})/\{\pm I\} \overset{\text{call}}{=} \mathrm{PSL}_2(\mathbf{C}).$$

This last group is, of course, the projective special linear group of 2-by-2 complex matrices. The upshot of all this is that to compose automorphisms, one may multiply their representing matrices of determinant 1, viewing any pair m and $-m$ of such matrices as equivalent. From now on we view $\mathrm{Aut}(\widehat{\mathbf{C}})$, $\mathrm{PGL}_2(\mathbf{C})$ and $\mathrm{PSL}_2(\mathbf{C})$ as identical, not merely isomorphic, and we freely use matrices to denote mappings. For example, $\begin{bmatrix} e^{i\alpha} & 0 \\ 0 & 1 \end{bmatrix} = \begin{bmatrix} e^{i\alpha/2} & 0 \\ 0 & e^{-i\alpha/2} \end{bmatrix}$ is the mapping $z \mapsto e^{i\alpha}z$.

Constructing the projective group of matrices $\mathrm{PGL}_2(\mathbf{C})$ from the linear group $\mathrm{GL}_2(\mathbf{C})$ is compatible with constructing projective space $\mathbf{P}^1(\mathbf{C})$ from affine space (without the origin) $\mathbf{C}^2 \setminus \mathbf{0}$, in the sense that the action of $\mathrm{GL}_2(\mathbf{C})$ as linear mappings of $\mathbf{C}^2$ descends to the natural action of $\mathrm{PGL}_2(\mathbf{C})$ on $\mathbf{P}^1(\mathbf{C})$. In other words, the following diagram commutes (Exercise 2.1.7).

$$\begin{array}{ccc} \mathrm{GL}_2(\mathbf{C}) \times \mathbf{C}^2 \setminus \mathbf{0} & \longrightarrow & \mathbf{C}^2 \setminus \mathbf{0} \\ {\scriptstyle \mathrm{P}\times P}\downarrow & & \downarrow{\scriptstyle P} \\ \mathrm{PGL}_2(\mathbf{C}) \times \mathbf{P}^1(\mathbf{C}) & \longrightarrow & \mathbf{P}^1(\mathbf{C}) \end{array} \tag{2.1.1}$$

This is just a fancy way of saying that constants pass through linear maps and are absorbed by projective classes.

Exercises

2.1.1. Show that for any $m, n \in \mathrm{GL}_2(\mathbf{C})$, $f_{mn} = f_m \circ f_n$, and that consequently $\{f_m : m \in \mathrm{GL}_2(\mathbf{C})\}$ forms a group.

2.1.2. Show that the kernel of the homomorphism $\mathrm{GL}_2(\mathbf{C}) \longrightarrow \mathrm{Aut}(\widehat{\mathbf{C}})$, where $m \mapsto f_m$, is $\mathbf{C}^*I = \{\lambda I : \lambda \in \mathbf{C}^*\}$.

2.1.3. Prove the First Isomorphism Theorem: A group homomorphism $\sigma : G \longrightarrow \tilde{G}$ with kernel K induces a natural isomorphism $\overline{\sigma} : G/K \xrightarrow{\sim} \sigma G$.

2.1.4. Verify that $K \triangleleft G$, $G = HK$ and $H \cap K = \{\pm I\}$ when $G = \mathrm{GL}_2(\mathbf{C})$, $H = \mathrm{SL}_2(\mathbf{C})$, and $K = \mathbf{C}^*I$. Do all three assertions remain valid if $\mathbf{C}$ is replaced by $\mathbf{R}$ throughout? Are $\mathrm{PGL}_2(\mathbf{R})$ and $\mathrm{PSL}_2(\mathbf{R})$ isomorphic?

2.1.5. Prove the Second Isomorphism Theorem: Let G be a group, $H \subset G$ a subgroup, and $K \triangleleft G$ a normal subgroup. Show that HK is a subgroup of G, that $K \triangleleft HK$, that $(H \cap K) \triangleleft H$, and that there is a natural isomorphism $H/(H \cap K) \xrightarrow{\sim} HK/K$.

2.1.6. Recall that the elements g, h of a group G are **conjugate** if $h = p^{-1}gp$ for some $p \in G$. Show that $m = \begin{bmatrix} 1 & 1 \\ 0 & 1 \end{bmatrix}$ and $n = \begin{bmatrix} -1 & 1 \\ 0 & -1 \end{bmatrix}$ are conjugate as elements of $\mathrm{PSL}_2(\mathbf{C})$ but not as elements of $\mathrm{SL}_2(\mathbf{C})$.

2.1.7. Verify the commutative diagram (2.1.1) by showing that for suitable matrices and vectors, the following diagram of elements commutes.

(2.1.2)
$$\begin{array}{ccc}
\left(\begin{bmatrix} a & b \\ c & d \end{bmatrix}, (z_1, z_2)\right) & \longrightarrow & (az_1 + bz_2, cz_1 + dz_2) \\
{\scriptstyle \mathrm{P}\times P}\downarrow & & \downarrow{\scriptstyle P} \\
\left(\overline{\begin{bmatrix} a & b \\ c & d \end{bmatrix}}, [z_1 : z_2]\right) & \longrightarrow & [az_1 + bz_2 : cz_1 + dz_2]
\end{array}$$

2. Rotations of the Riemann sphere

A **rotation** of the sphere S^2 is a map $r = r_{p,\alpha}$ described by spinning the sphere (actually, spinning the ambient space $\mathbf{R}^3$) about the line through the origin and the point $p \in S^2$, counterclockwise through angle α looking at p from outside the sphere. (See Figure 2.2.1.)

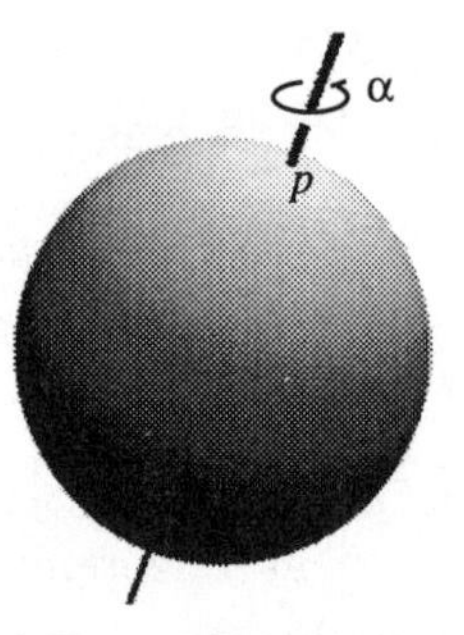

Figure 2.2.1. The rotation $r_{p,\alpha}$

Thus r is the linear map that fixes p and rotates planes orthogonal to p through angle α. Let q be a unit vector orthogonal to p. Then the matrix of r is (viewing p and q as column vectors)

$$m_r = \begin{bmatrix} p & q & p \times q \end{bmatrix} \begin{bmatrix} 1 & 0 & 0 \\ 0 & \cos\alpha & -\sin\alpha \\ 0 & \sin\alpha & \cos\alpha \end{bmatrix} \begin{bmatrix} p & q & p \times q \end{bmatrix}^{-1}.$$

(Exercise 2.2.1 asks for the proof.) The set $\mathrm{Rot}(S^2)$ of such rotations forms a group, most naturally viewed as a subgroup of $\mathrm{GL}_3(\mathbf{R})$. Showing this requires some linear algebra.

Recall that if $m \in \mathrm{M}_3(\mathbf{R})$, meaning m is a 3-by-3 real matrix, then its **transpose** m^t is obtained by flipping about the diagonal; thus $m^t_{ij} = m_{ji}$

for $i, j = 1, 2, 3$, where the subscripts specify matrix entries. The transpose is characterized by the more convenient condition

$$\langle mx, y \rangle = \langle x, m^t y \rangle \quad \text{for all } x, y \in \mathbf{R}^3,$$

where $\langle\ ,\ \rangle$ is the usual inner product $\langle x, y \rangle = \sum x_i y_i$ (Exercise 2.2.2). The matrix m is **orthogonal** if $m^t m = I$ or equivalently (Exercise 2.2.3) if m preserves inner products,

$$\langle mx, my \rangle = \langle x, y \rangle \quad \text{for all } x, y \in \mathbf{R}^3.$$

The orthogonal matrices form a group $\mathrm{O}_3(\mathbf{R}) \subset \mathrm{GL}_3(\mathbf{R})$, and the special orthogonal matrices

$$\mathrm{SO}_3(\mathbf{R}) = \{m \in \mathrm{O}_3(\mathbf{R}) : \det m = 1\}$$

form a subgroup of index 2 (Exercise 2.2.4). With these facts in place it is not hard to prove (Exercise 2.2.5) that $\mathrm{Rot}(S^2)$ forms a group, and

(2.2.1) THEOREM. *As a subgroup of* $\mathrm{GL}_3(\mathbf{R})$, $\mathrm{Rot}(S^2) = \mathrm{SO}_3(\mathbf{R})$.

A **rotation** of the Riemann sphere $\widehat{\mathbf{C}}$ is a map $f : \widehat{\mathbf{C}} \longrightarrow \widehat{\mathbf{C}}$ corresponding under stereographic projection to a true rotation r of the sphere S^2. In other words, the following diagram commutes.

$$\begin{array}{ccc} S^2 & \xrightarrow{\ r\ } & S^2 \\ {\scriptstyle\pi}\downarrow & & \downarrow{\scriptstyle\pi} \\ \widehat{\mathbf{C}} & \xrightarrow{\ f\ } & \widehat{\mathbf{C}} \end{array}$$

Let $\mathrm{Rot}(\widehat{\mathbf{C}})$ denote these rotations. Since $\mathrm{Rot}(S^2)$ forms a group, $\mathrm{Rot}(\widehat{\mathbf{C}})$ forms an isomorphic group under $r \mapsto \pi \circ r \circ \pi^{-1}$. Since any rotation r is conformal on S^2, the corresponding bijection f is conformal on $\widehat{\mathbf{C}}$ and is therefore an automorphism, so $\mathrm{Rot}(\widehat{\mathbf{C}})$ is a subgroup of $\mathrm{Aut}(\widehat{\mathbf{C}})$. With some more linear algebra we can describe $\mathrm{Rot}(\widehat{\mathbf{C}})$ explicitly as a subgroup of $\mathrm{PSL}_2(\mathbf{C})$.

If $m \in \mathrm{M}_2(\mathbf{C})$ is a 2-by-2 complex matrix then its **adjoint** $m^* = \overline{m}^t$ (the overbar means complex conjugation, thus $m^*_{ij} = \overline{m_{ji}}$ for $i, j = 1, 2$) is characterized by the condition

$$\langle mx, y \rangle = \langle x, m^* y \rangle \quad \text{for all } x, y \in \mathbf{C}^2,$$

where now $\langle\ ,\ \rangle$ is the complex inner product $\langle x, y \rangle = \sum \overline{x_i} y_i$. (See Exercises 2.2.7 through 2.2.9 for proofs of the statements in this paragraph.) The role of the adjoint in the algebra of complex matrices is analogous to the role of the conjugate in the algebra of complex numbers. The matrix

u is **unitary** if $u^*u = I$, which generalizes the unit complex numbers, or equivalently

$$\langle ux, uy\rangle = \langle x, y\rangle \quad \text{for all } x, y \in \mathbf{C}^2.$$

The unitary matrices form a group $\mathrm{U}_2(\mathbf{C})$. The special unitary matrices

$$\mathrm{SU}_2(\mathbf{C}) = \{u \in \mathrm{U}_2(\mathbf{C}) : \det u = 1\}$$

form a subgroup. A matrix is special unitary if and only if it takes the form $\begin{bmatrix} a & b \\ -\bar{b} & \bar{a} \end{bmatrix}$ with $|a|^2 + |b|^2 = 1$. The projective unitary group is

$$\mathrm{PU}_2(\mathbf{C}) = \mathrm{U}_2(\mathbf{C})/(\mathrm{U}_2(\mathbf{C}) \cap \mathbf{C}^* I),$$

and the projective special unitary group is

$$\mathrm{PSU}_2(\mathbf{C}) = \mathrm{SU}_2(\mathbf{C})/(\mathrm{SU}_2(\mathbf{C}) \cap \mathbf{C}^* I) = \mathrm{SU}_2(\mathbf{C})/\{\pm I\}.$$

As in Section 2.1 with the general and special linear groups, $\mathrm{PU}_2(\mathbf{C}) \cong \mathrm{PSU}_2(\mathbf{C})$, and $\mathrm{PSU}_2(\mathbf{C})$ is more convenient since its elements are two-element cosets $\{\pm u\}$.

(2.2.2) THEOREM. *As a subgroup of* $\mathrm{PSL}_2(\mathbf{C})$, $\mathrm{Rot}(\widehat{\mathbf{C}}) = \mathrm{PSU}_2(\mathbf{C})$.

With Theorem 2.2.1 this says that $\mathrm{PSU}_2(\mathbf{C}) \cong \mathrm{SO}_3(\mathbf{R})$. See Exercise 2.2.10 for an elegant proof, which shows that $\mathrm{Rot}(\widehat{\mathbf{C}})$ is a group without reference to $\mathrm{SO}_3(\mathbf{R})$. The next result says how to compute explicitly in $\mathrm{PSU}_2(\mathbf{C})$ while thinking of $\mathrm{Rot}(S^2)$. For any rotation $r_{p,\alpha}$ of S^2, let $f_{\pi(p),\alpha}$ denote the corresponding rotation of $\widehat{\mathbf{C}}$.

(2.2.3) THEOREM. *Let* $p = (p_1, p_2, p_3) \in S^2$ *and let* $\alpha \in \mathbf{R}$. *Then*

$$f_{\pi(p),\alpha} = \begin{bmatrix} \cos\frac{\alpha}{2} + ip_3\sin\frac{\alpha}{2} & -p_2\sin\frac{\alpha}{2} + ip_1\sin\frac{\alpha}{2} \\ p_2\sin\frac{\alpha}{2} + ip_1\sin\frac{\alpha}{2} & \cos\frac{\alpha}{2} - ip_3\sin\frac{\alpha}{2} \end{bmatrix}.$$

PROOF. Either by geometry or by a calculation using the commutative diagram after Theorem 2.2.1, $r_{\mathbf{n},\alpha}$ induces the automorphism $f_{\infty,\alpha}(z) = e^{i\alpha}z$ of $\widehat{\mathbf{C}}$, i.e., $f_{\infty,\alpha} = \begin{bmatrix} e^{i\alpha/2} & 0 \\ 0 & e^{-i\alpha/2} \end{bmatrix}$ (Exercise 2.2.11).

Next consider the rotation $r_{(0,1,0),\phi}$ of S^2 counterclockwise about the positive x_2-axis through angle ϕ. We will find the corresponding rotation $f_{i,\phi}$ of $\widehat{\mathbf{C}}$.

A rotation r of S^2 takes $(0,1,0)$ to $\mathbf{n}$ and $(0,-1,0)$ to $\mathbf{s}$; the corresponding rotation f of $\widehat{\mathbf{C}}$ takes i to ∞ and $-i$ to 0, so it has the form $f(z) = k(z+i)/(z-i)$ for some nonzero constant k. Since $r_{(0,1,0),\phi} = r^{-1} \circ r_{\mathbf{n},\phi} \circ r$, the corresponding result in $\text{Rot}(\widehat{\mathbf{C}})$ is $f_{i,\phi} = f^{-1} \circ f_{\infty,\phi} \circ f$, or $f \circ f_{i,\phi} = f_{\infty,\phi} \circ f$. Thus for all $z \in \widehat{\mathbf{C}}$,

$$k \cdot \frac{f_{i,\phi}(z) + i}{f_{i,\phi}(z) - i} = e^{i\phi} k \cdot \frac{z+i}{z-i}.$$

The k cancels, leaving $e^{-i\phi/2}(f_{i,\phi}(z) + i)(z - i) = e^{i\phi/2}(f_{i,\phi}(z) - i)(z + i)$, and some algebra gives $f_{i,\phi} = \begin{bmatrix} \cos\frac{\phi}{2} & -\sin\frac{\phi}{2} \\ \sin\frac{\phi}{2} & \cos\frac{\phi}{2} \end{bmatrix}$.

Now let the point $p \in S^2$ have spherical coordinates $(1, \theta, \phi)$, meaning $\cos\theta = p_1/\sqrt{p_1^2 + p_2^2}$, $\sin\theta = p_2/\sqrt{p_1^2 + p_2^2}$, $\cos\phi = p_3$, and $\sin\phi = \sqrt{p_1^2 + p_2^2}$. (See Figure 2.2.2.)

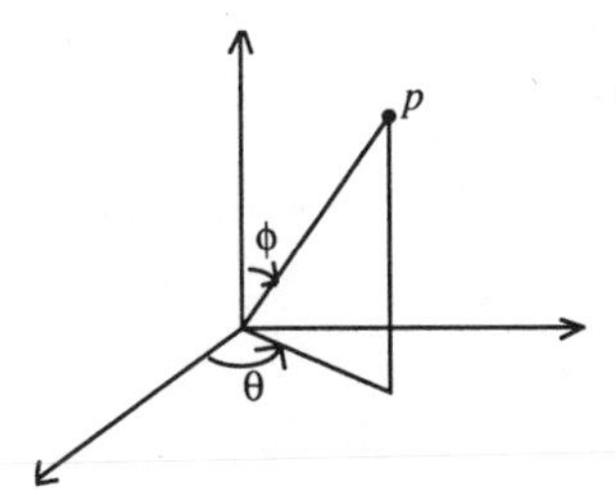

Figure 2.2.2. Spherical coordinates

To carry out $r_{p,\alpha}$, move p to the north pole via rotations about the north pole and $(0,1,0)$, rotate about the north pole by α, and restore p; to wit,

$$r_{p,\alpha} = r_{\mathbf{n},\theta} \circ r_{(0,1,0),\phi} \circ r_{\mathbf{n},\alpha} \circ r_{(0,1,0),-\phi} \circ r_{\mathbf{n},-\theta}.$$

The corresponding rotation of $\widehat{\mathbf{C}}$ is

$$f_{\pi(p),\alpha} = \begin{bmatrix} e^{i\theta/2} & 0 \\ 0 & e^{-i\theta/2} \end{bmatrix} \begin{bmatrix} \cos\frac{\phi}{2} & -\sin\frac{\phi}{2} \\ \sin\frac{\phi}{2} & \cos\frac{\phi}{2} \end{bmatrix} \begin{bmatrix} e^{i\alpha/2} & 0 \\ 0 & e^{-i\alpha/2} \end{bmatrix}$$
$$\cdot \begin{bmatrix} \cos\frac{\phi}{2} & \sin\frac{\phi}{2} \\ -\sin\frac{\phi}{2} & \cos\frac{\phi}{2} \end{bmatrix} \begin{bmatrix} e^{-i\theta/2} & 0 \\ 0 & e^{i\theta/2} \end{bmatrix}.$$

Multiplying this out and using a little trigonometry gives the result. □

Any special unitary matrix takes the form $\begin{bmatrix} d+ic & -b+ia \\ b+ia & d-ic \end{bmatrix}$ with $a, b, c, d \in \mathbf{R}$ and $a^2 + b^2 + c^2 + d^2 = 1$, and the conditions $a = p_1 \sin(\alpha/2)$, $b = p_2 \sin(\alpha/2)$, $c = p_3 \sin(\alpha/2)$, $d = \cos(\alpha/2)$ determine values of $p \in S^2$ and $\alpha \in \mathbf{R}$ for which the matrix represents $f_{\pi(p),\alpha}$. (This is Exercise 2.2.13.) Going from a, b, c, d to p, α and from there to the special orthogonal matrix representing $r_{p,\alpha}$ gives the isomorphism $\text{PSU}_2(\mathbf{C}) \xrightarrow{\sim} \text{SO}_3(\mathbf{R})$.

The spherical rotations $\text{Rot}(S^2)$ are convenient for geometric intuition, while $\text{PSU}_2(\mathbf{C})$ is well suited for computation.

Exercises

2.2.1. Explain why the rotation $r = r_{p,\alpha}$ has matrix m_r as given in the text.

2.2.2. Show that the definition of transpose is equivalent to its inner product characterization. That is, show that indeed $\langle mx, y\rangle = \langle x, m^t y\rangle$ for all $x, y \in \mathbf{R}^3$, and conversely if $\langle mx, y\rangle = \langle x, ny\rangle$ for all $x, y \in \mathbf{R}^3$ then $n = m^t$. (Compute $\langle me_i, e_j\rangle$ and $\langle e_i, ne_j\rangle$ where e_i and e_j are standard basis vectors.) Use the inner product characterization of transpose to show that $(mn)^t = n^t m^t$ for all $m, n \in \mathrm{M}_3(\mathbf{R})$, that $I^t = I$, and that $(m^{-1})^t = (m^t)^{-1}$ for all $m \in \mathrm{GL}_3(\mathbf{R})$.

2.2.3. Prove the following useful lemma: For $m, n \in \mathrm{M}_3(\mathbf{R})$,

$$\langle x, my\rangle = \langle x, ny\rangle \text{ for all } x, y \in \mathbf{R}^3 \quad \Longleftrightarrow \quad m = n.$$

(Hint: subtracting the right side in each equality reduces this to the case $n = 0$.) Use the lemma to show that the two characterizations of orthogonal are equivalent.

2.2.4. The condition $m^t m = I$ shows that orthogonal matrices are invertible and $|\det m| = 1$ if m is orthogonal. Use the inner product characterization of orthogonal to show that $\mathrm{O}_3(\mathbf{R})$ contains the identity, is closed under multiplication and is closed under inversion, so it forms a subgroup of $\mathrm{GL}_3(\mathbf{R})$. Use the homomorphism $\det : \mathrm{O}_3(\mathbf{R}) \longrightarrow \{x \in \mathbf{R} : |x| = 1\}$ to show that $\mathrm{SO}_3(\mathbf{R})$ is a subgroup of $\mathrm{O}_3(\mathbf{R})$ of index 2.

2.2.5. This exercise shows that $\mathrm{Rot}(S^2)$ is a group isomorphic to $\mathrm{SO}_3(\mathbf{R})$. Fill in details as necessary. Given a rotation $r = r_{p,\alpha}$, its matrix as given in the text,

$$m_r = \begin{bmatrix} p & q & p \times q \end{bmatrix} \begin{bmatrix} 1 & 0 & 0 \\ 0 & \cos\alpha & -\sin\alpha \\ 0 & \sin\alpha & \cos\alpha \end{bmatrix} \begin{bmatrix} p & q & p \times q \end{bmatrix}^{-1},$$

is special orthogonal. On the other hand, take any special orthogonal matrix m. Since 3 is odd, m has a real eigenvalue λ. Any real eigenvalue λ with eigenvector p satisfies

$$\lambda^2 \langle p, p\rangle = \langle \lambda p, \lambda p\rangle = \langle mp, mp\rangle = \langle p, p\rangle,$$

i.e., $\lambda = \pm 1$. Since $\det m = 1$, and the determinant is the product of the eigenvalues, and any imaginary eigenvalues occur in conjugate pairs, m in fact has 1 for an eigenvalue with unit eigenvector p. Take any nonzero vector q perpendicular to p. Some rotation $r = r_{p,\alpha}$ takes

q to mq and has matrix $m_r \in \mathrm{SO}_3(\mathbf{R})$. Thus the matrix

$$m_r^{-1} m$$

lies in $\mathrm{SO}_3(\mathbf{R})$ and fixes both p and q. It is therefore the identity, showing $m = m_r$ is a rotation matrix.

2.2.6. A **rigid motion** of $\mathbf{R}^3$ is a mapping $R : \mathbf{R}^3 \longrightarrow \mathbf{R}^3$ such that

$$\langle R(y) - R(x), R(z) - R(x) \rangle = \langle y - x, z - x \rangle \qquad \text{for all } x, y, z \in \mathbf{R}^3.$$

This exercise proves that the rigid motions are precisely the affine mappings $R(x) = Ax + b$ where $A \in \mathrm{O}_3(\mathbf{R})$ and $b \in \mathbf{R}^3$.
(a) Prove that any such affine mapping is rigid.
(b) Now let $R : \mathbf{R}^3 \longrightarrow \mathbf{R}^3$ be rigid. Show that R is a bijection. Let $b = R(\mathbf{0})$ and define $S : \mathbf{R}^3 \longrightarrow \mathbf{R}^3$ by $S(x) = R(x) - b$. Thus $S(\mathbf{0}) = \mathbf{0}$. Apply the definition of rigidity to S with $x = \mathbf{0}$ to get $\langle S(y), S(z) \rangle = \langle y, z \rangle$ for all $y, z \in \mathbf{R}^3$. Show that for any $x, y, z \in \mathbf{R}^3$, $\langle S(x + y), S(z) \rangle = \langle S(x) + S(y), S(z) \rangle$, so S preserves addition. Similarly, show that S preserves scalar multiplication, so S is linear. Therefore $S(x) = Ax$ for some matrix $A \in \mathrm{M}_3\mathbf{R}$. Finally, $\langle Ax, Ay \rangle = \langle S(x), S(y) \rangle = \langle x, y \rangle$, so $A \in \mathrm{O}_3(\mathbf{R})$.

2.2.7. Repeat Exercises 2.2.2, 2.2.3, and 2.2.4 with transpose and orthogonal in $\mathrm{M}_3(\mathbf{R})$ replaced by adjoint and unitary in $\mathrm{M}_2(\mathbf{C})$. Note that the index of $\mathrm{SU}_2(\mathbf{C})$ in $\mathrm{U}_2(\mathbf{C})$ is infinite.

2.2.8. Show that a matrix $m \in \mathrm{M}_2(\mathbf{C})$ is special unitary if and only if it takes the form $m = \begin{bmatrix} a & b \\ -\overline{b} & \overline{a} \end{bmatrix}$ with $|a|^2 + |b|^2 = 1$.

2.2.9. Show that $\mathrm{U}_2(\mathbf{C}) \cap \mathbf{C}^* I = \{\lambda I : |\lambda| = 1\}$. If $u \in \mathrm{U}_2(\mathbf{C})$, show that $\lambda u \in \mathrm{U}_2(\mathbf{C})$ if and only if $|\lambda| = 1$. Show that $\mathrm{U}_2(\mathbf{C}) = \mathrm{SU}_2(\mathbf{C}) \cdot (\mathrm{U}_2(\mathbf{C}) \cap \mathbf{C}^* I)$. Show that $\mathrm{U}_2(\mathbf{C}) \cap \mathbf{C}^* I \triangleleft \mathrm{U}_2(\mathbf{C})$. Show that $\mathrm{SU}_2(\mathbf{C}) \cap \mathbf{C}^* I = \{\pm I\}$. Show that $\mathrm{PU}_2(\mathbf{C}) \cong \mathrm{PSU}_2(\mathbf{C})$, and the elements of $\mathrm{PSU}_2(\mathbf{C})$ are two-element cosets $\{\pm u\}$.

2.2.10. Here is a proof of Theorem 2.2.2. We show first that any rotation lies in $\mathrm{PSU}_2(\mathbf{C})$, second that any element of $\mathrm{PSU}_2(\mathbf{C})$ is a rotation. Show that if the antipodal pair $p, -p \in S^2 \setminus \{\mathbf{n}, \mathbf{s}\}$ have stereographic images $z, z^* \in \mathbf{C}$, then $z^* = -1/\overline{z}$, where the overbar is complex conjugation.

Now let $f = \begin{bmatrix} a & b \\ c & d \end{bmatrix}$ (normalized to determinant 1) be a rotation of $\widehat{\mathbf{C}}$ induced by a rotation r of S^2. Since r takes antipodal pairs to antipodal pairs, f must satisfy the corresponding relation $f(z^*) =$

$f(z)^*$ for all $z \in \mathbf{C} \setminus \{0\}$. Show that this condition is that for some $\lambda \in \mathbf{C}^*$,

$$d = \lambda\bar{a}, \quad a = \lambda\bar{d}, \quad c = -\lambda\bar{b}, \quad b = -\lambda\bar{c}.$$

Use these relations and $ad - bc = 1$ to show that $\lambda = 1$ and therefore $f \in \mathrm{PSU}_2(\mathbf{C})$.

For the converse, let $f = \begin{bmatrix} a & b \\ -\bar{b} & \bar{a} \end{bmatrix} \in \mathrm{PSU}_2(\mathbf{C})$. Show that if $f(0) = 0$ then $f(z) = e^{i\alpha}z$ for some α, so f is a rotation. If $f(0) = z \neq 0$ then some rotation $f_z \in \mathrm{Rot}(\widehat{\mathbf{C}}) \subset \mathrm{PSU}_2(\mathbf{C})$ takes z to 0, so $g = f_z^{-1} \circ f \in \mathrm{PSU}_2(\mathbf{C})$ fixes 0 and is thus a rotation. Therefore $f = f_z \circ g$ is also a rotation and the proof is complete.

2.2.11. Show that the rotation $r_{\mathbf{n},\alpha}$ of S^2 induces the rotation $f_{\infty,\alpha}(z) = e^{i\alpha}z$ of $\widehat{\mathbf{C}}$.

2.2.12. Carry out the two calculations omitted from the proof of Theorem 2.2.3.

2.2.13. Show that the conditions $a = p_1 \sin(\alpha/2)$, $b = p_2 \sin(\alpha/2)$, $c = p_3 \sin(\alpha/2)$, $d = \cos(\alpha/2)$ determine values of $p \in S^2$ and $\alpha \in \mathbf{R}$.

3. Finite automorphism groups and rotation groups

The finite automorphism groups of the sphere, i.e., the finite subgroups of $\mathrm{Aut}(\widehat{\mathbf{C}})$, collectively form too unwieldy a set to classify conveniently. In general, if a classification problem is too hard, one simplifies it by putting an equivalence relation on the objects to be classified and then classifying equivalence classes instead. Recall that two subgroups G_1 and G_2 of a group G are **conjugate** if $G_2 = gG_1g^{-1}$ for some $g \in G$. The next theorem shows that in the case of finite automorphism groups of the sphere, conjugacy is a happy choice of equivalence because each conjugacy class of such groups is represented by a rotation group, meaning a subgroup of $\mathrm{Rot}(\widehat{\mathbf{C}})$. Classifying the finite rotation groups in turn is an essentially geometrical process which we will carry out in Section 2.6.

(2.3.1) THEOREM. *Any finite automorphism group of the sphere is conjugate to a rotation group.*

Recall that $\mathrm{Aut}(\widehat{\mathbf{C}})$ is identified with $\mathrm{PSL}_2(\mathbf{C})$ and that under this identification $\mathrm{Rot}(\widehat{\mathbf{C}}) = \mathrm{PSU}_2(\mathbf{C})$. Thus, in projective matrix terms the theorem is that any finite subgroup of $\mathrm{PSL}_2(\mathbf{C})$ is conjugate to a subgroup of $\mathrm{PSU}_2(\mathbf{C})$. This assertion reduces easily to a nonprojective counterpart.

(2.3.2) PROPOSITION. *Any finite subgroup G of* $\mathrm{GL}_2(\mathbf{C})$ *is conjugate to a subgroup of* $\mathrm{U}_2(\mathbf{C})$.

To prove Theorem 2.3.1 from the proposition, take any finite group of automorphisms $\Gamma \subset \mathrm{PSL}_2(\mathbf{C})$. This lifts to a finite group $\Gamma' \subset \mathrm{SL}_2(\mathbf{C})$ with twice as many elements as Γ (see Exercise 2.3.1), which by the proposition is conjugate to a subgroup of $\mathrm{SU}_2(\mathbf{C})$. The matrix $p \in \mathrm{GL}_2(\mathbf{C})$ such that $p\Gamma'p^{-1} \subset \mathrm{SU}_2(\mathbf{C})$ may be taken to have determinant 1, so it defines an element of $\mathrm{PSL}_2(\mathbf{C})$, also called p, such that $p\Gamma p^{-1} \subset \mathrm{PSU}_2(\mathbf{C})$, QED.

Before proving Proposition 2.3.2 we review some inner product theory. The usual inner product on $\mathbf{C}^2$, $\langle x, y\rangle = \sum \overline{x_i} y_i$, is **linear in the second argument**, **skew symmetric**, and **positive definite**, meaning

$$\begin{aligned}
\langle x, ay + y'\rangle &= a\langle x, y\rangle + \langle x, y'\rangle && \text{for all } a \in \mathbf{C} \text{ and } x, y, y' \in \mathbf{C}^2,\\
\langle y, x\rangle &= \overline{\langle x, y\rangle} && \text{for all } x, y \in \mathbf{C}^2,\\
\langle x, x\rangle &\geq 0 && \text{for all } x \in \mathbf{C}^2,\\
& && \quad \text{with equality if and only if } x = 0.
\end{aligned}$$

These are called the **standard properties** of the inner product (Exercise 2.3.2). By referring to these rather than to the coordinate-dependent definition, proofs in inner product theory apply more broadly to any inner product satisfying the same three properties. For example, the first two properties show that the inner product is **conjugate linear** in the first argument, meaning

$$\langle ax + x', y\rangle = \overline{a}\langle x, y\rangle + \langle x', y\rangle \quad \text{for all } a \in \mathbf{C} \text{ and } x, x', y \in \mathbf{C}^2.$$

(This is Exercise 2.3.3.)

Any inner product $[\ ,\ ]$ on $\mathbf{C}^2$ satisfying the standard properties defines a matrix $m \in \mathrm{M}_2(\mathbf{C})$ with entries $m_{ij} = [e_i, e_j]$ for $i, j = 1, 2$, where e_i, e_j are the ith and jth standard basis vectors. Linearity and conjugate linearity show that the matrix m characterizes the new inner product in terms of the usual one, $[x, y] = \langle x, my\rangle = x^*my$, where x^* means to transpose x into a row vector and conjugate its entries (Exercise 2.3.4). Skew symmetry implies that m is **Hermitian** (or **self-adjoint**), meaning $m^* = m$; positive definiteness says that m is **positive**, meaning $x^*mx > 0$ for all nonzero $x \in \mathbf{C}^2$ (Exercise 2.3.5). Conversely, given a Hermitian positive matrix m, the inner product $[x, y] = \langle x, my\rangle$ satisfies the standard properties (Exercise 2.3.6). Thus inner products are characterized by Hermitian positive matrices.

Every Hermitian matrix has real eigenvalues and is diagonalized by a unitary matrix. Indeed, if m is Hermitian with eigenvalue λ and corresponding unit eigenvector x then

$$\lambda = \lambda\langle x, x\rangle = \langle x, \lambda x\rangle = \langle x, mx\rangle = \langle x, m^*x\rangle = \langle mx, x\rangle = \langle \lambda x, x\rangle = \overline{\lambda}\langle x, x\rangle$$
$$= \overline{\lambda},$$

so $\lambda \in \mathbf{R}$. And if y is a unit vector orthogonal to x, meaning $\langle x, y\rangle = 0$, then

$$\langle x, my\rangle = \langle x, m^*y\rangle = \langle mx, y\rangle = \langle \lambda x, y\rangle = \overline{\lambda}\langle x, y\rangle = 0,$$

so also my is orthogonal to x. Since $\dim(\mathbf{C}^2) = 2$, this says that $my = \mu y$ for some μ. Thus the matrix $u = [x \quad y] \in \mathrm{U}_2(\mathbf{C})$ satisfies $u^{-1} = u^*$ and

$$u^*mu = \begin{bmatrix} \lambda & 0 \\ 0 & \mu \end{bmatrix}.$$

(Exercise 2.3.7 asks for details and the generalization to the n-by-n case.) This diagonalization shows that a Hermitian matrix m is positive if and only if all of its eigenvalues are positive reals, which in turn is equivalent to the condition $m = p^2$ where p is also Hermitian positive (Exercise 2.3.8). Thus any inner product satisfying the standard properties takes the form

$$[x, y] = \langle x, my\rangle = \langle x, p^*py\rangle = \langle px, py\rangle \quad \text{for some Hermitian positive } p.$$

Now it is easy to show that any finite subgroup G of $\mathrm{GL}_2(\mathbf{C})$ is conjugate to a subgroup of $\mathrm{U}_2(\mathbf{C})$.

Proof of Proposition 2.3.2. Define an inner product $[\ ,\ ]$ on $\mathbf{C}^2$ by averaging the usual one over G,

$$[x, y] = \frac{1}{|G|}\sum_{\gamma\in G}\langle \gamma x, \gamma y\rangle \qquad \text{for all } x, y \in \mathbf{C}^2.$$

This new inner product is G-invariant, meaning $[gx, gy] = [x, y]$ for all $g \in G$ and $x, y \in \mathbf{C}^2$ (Exercise 2.3.9). It satisfies the usual inner product properties, so $[x, y] = \langle px, py\rangle$ (and equivalently, $\langle x, y\rangle = [p^{-1}x, p^{-1}y]$) for some Hermitian positive p and all $x, y \in \mathbf{C}^2$. Therefore, for any $g \in G$,

$$\langle x, y\rangle = [p^{-1}x, p^{-1}y] = [gp^{-1}x, gp^{-1}y] = \langle pgp^{-1}x, pgp^{-1}y\rangle \quad \text{for all } x, y \in \mathbf{C}^2$$

and so pgp^{-1} is unitary. □

The proposition generalizes immediately to n-by-n matrices, and proves that any complex representation of a finite group is similar to a unitary representation.

Actually, a stronger result than Theorem 2.3.1 holds. An element γ of $\mathrm{PSL}_2(\mathbf{C})$ is called **elliptic** if some $p_\gamma \in \mathrm{PSL}_2(\mathbf{C})$ conjugates it to a rotation, i.e., $p_\gamma \gamma p_\gamma^{-1} = r_\gamma$ for some $r_\gamma \in \mathrm{PSU}_2(\mathbf{C})$. Each $\gamma \in \mathrm{PSL}_2(\mathbf{C})$ of finite order is elliptic, but not all elliptic elements have finite order. Thus finite subgroups of $\mathrm{PSL}_2(\mathbf{C})$ consist of elliptic elements, but a subgroup of $\mathrm{PSL}_2(\mathbf{C})$ whose elements are elliptic need not be finite. The stronger result is that if G is a subgroup of $\mathrm{PSL}_2(\mathbf{C})$ whose elements are elliptic, then G is conjugate to a subgroup of $\mathrm{PSU}_2(\mathbf{C})$—that is, a single p conjugates all $\gamma \in G$ to rotations. For a proof, see [Jo-Si] or their reference, Lyndon and Ullman [Ly-Ul].

Exercises

2.3.1. Let $f : G' \longrightarrow G$ be a surjective group homomorphism. Suppose $H \subset G$ is a subgroup. Define $H' = f^{-1}(H) \subset G'$. Show that H' is a subgroup of G'. If $\ker f$ and H are finite, show that $|H'| = |\ker f||H|$.

2.3.2. Show that the usual inner product on $\mathbf{C}^2$ has the standard properties.

2.3.3. Show that any inner product with the standard properties is conjugate linear in the first argument.

2.3.4. Given any inner product $[x, y]$ on $\mathbf{C}^2$ satisfying the standard properties, define a matrix $m \in \mathrm{M}_2(\mathbf{C})$ by $m_{ij} = [e_i, e_j]$ for $i, j = 1, 2$. Show that $[x, y] = \langle x, my \rangle = x^* my$.

2.3.5. Show that the matrix m associated with an inner product $[x, y]$ satisfying the usual properties is Hermitian positive.

2.3.6. Given a Hermitian positive matrix m, show that the inner product $[x, y] = \langle x, my \rangle$ satisfies the standard properties.

2.3.7. Supply details as necessary in the proof that a Hermitian matrix m is diagonalized by a unitary matrix. Generalize to the n-by-n case. (Show that the subspace of $\mathbf{C}^n$ orthogonal to x is preserved under m, and use induction.)

2.3.8. Show that a Hermitian matrix m is positive if and only if all of its eigenvalues are positive reals, which in turn is equivalent to the condition $m = p^2$ where p is also Hermitian positive.

2.3.9. Show that the inner product in the proof of Proposition 2.3.2 is G-invariant.

2.3.10. Show that all elements of $\mathrm{PSL}_2(\mathbf{C})$ of finite order are elliptic by inspecting the Jordan forms of their matrix representatives. (Recall that $(p^{-1}Jp)^n = p^{-1}J^n p$.) When does an elliptic element have finite order?

4. Group actions

To continue the classification program, we need to connect the geometry of the sphere to the algebra of its rotation group. The next salvo of terminology is for analyzing the interplay between a group and a set it permutes.

An **action** of a group G on a set S is a map

$$G \times S \longrightarrow S,$$

notated by the juxtaposition $(g, s) \mapsto gs$, such that $1_G s = s$ for all $s \in S$ and $(g_1 g_2)s = g_1(g_2 s)$ for all $g_1, g_2 \in G$ and $s \in S$. The points $s_1, s_2 \in S$ are **congruent modulo** G if $s_2 = g s_1$ for some $g \in G$. This is an equivalence relation (Exercise 2.4.1). The **orbit** of a point $s \in S$ under the action is the set of points congruent to s modulo G, $\mathcal{O}_s = \{gs : g \in G\}$. Orbits are equal or disjoint since congruence modulo G is an equivalence on S (Exercise 2.4.1 again). The action is **transitive** if all points of S are congruent modulo G, i.e., there is only one orbit. The **length** of an orbit is its cardinality $|\mathcal{O}_s|$. The **stabilizer** of s is $\text{stab}(s) = \{g \in G : gs = s\}$, a subgroup of G. The **Orbit-Stabilizer Theorem** says that $|G| = |\mathcal{O}_s| \cdot |\text{stab}(s)|$. ($G$ might be infinite, in which case so is at least one of $\mathcal{O}_s$ and $\text{stab}(s)$.) If $\tilde{S}$ is a subset of S, a **symmetry** of $\tilde{S}$ is a group element $g \in G$ such that $g\tilde{S} = \tilde{S}$. The symmetries of $\tilde{S}$ form a subgroup $\text{sym}(\tilde{S})$ of G, called the **symmetry group of** $\tilde{S}$; in particular, $\text{sym}(\{s\}) = \text{stab}(s)$. (Exercise 2.4.3 asks for proofs of these statements.)

The **kernel** of the action of G on S is $K = \{g \in G : gs = s \text{ for all } s \in S\}$, a normal subgroup of G. If K is trivial, the action is called **faithful** because each group element acts differently, meaning no information is lost going from the group to its action. If S is a finite set of n elements, G/K is isomorphic to a subgroup of S_n, the group of symmetries on $\{1, \ldots, n\}$; so in particular, if the action is faithful, then G itself is isomorphic to a subgroup of S_n. (See Exercise 2.4.4.)

In the next set of exercises, the group G acts on the set S.

Exercises

2.4.1. Prove that congruence modulo G is an equivalence relation, and therefore orbits are equal or disjoint.

2.4.2. Let T be the set of triangles in the complex plane. What subsets of T are the orbits under the actions of the following groups? (a) $G_1 = \left\{\begin{bmatrix} a & b \\ 0 & 1 \end{bmatrix} \in \text{PGL}_2(\mathbf{C})\right\}$, (b) $G_2 = \left\{\begin{bmatrix} a & b \\ 0 & 1 \end{bmatrix} \in \text{PGL}_2(\mathbf{C}) : |a| = 1\right\}$,

(c) G_3, the group generated by G_1 and complex conjugation, (d) G_4, the group generated by G_2 and complex conjugation.

2.4.3. For any subset $\tilde{S}$ of S, prove that $\mathrm{sym}(\tilde{S})$ is a subgroup of G, so in particular $\mathrm{stab}(s)$ is a subgroup for each $s \in S$. Prove the Orbit-Stabilizer Theorem.

2.4.4. Let the action of G on S have kernel K. Show that K is a normal subgroup of G. Show that if K is trivial then each element of G acts differently on S. If S is a finite set of n elements, show that G/K is isomorphic to a subgroup of S_n, so if the action is faithful, then G itself is isomorphic to a subgroup of S_n.

2.4.5. Suppose $H \subset G$ is a subgroup, which inherits an action on S from G, and $\mathcal{O}$ is an H-orbit containing the point s. Show that for any $\gamma \in G$,

$$\gamma\mathcal{O} \cap \mathcal{O} \neq \emptyset \qquad \Longleftrightarrow \qquad \gamma \in H \cdot \mathrm{stab}(s) \cdot H.$$

2.4.6. Suppose S_1 and S_2 are subsets of S with symmetry groups G_1 and G_2, subgroups of the symmetry group G of S. Extending the terminology in the text, the sets S_1 and S_2 are **congruent modulo** G, written $S_1 \equiv S_2 \pmod{G}$, if $S_2 = gS_1$ for some $g \in G$. Recall that G_1 and G_2 are conjugate in G, written $G_1 \sim G_2 \pmod{G}$, if $G_2 = gG_1g^{-1}$ for some $g \in G$. Show that the geometric condition that S_1 and S_2 are congruent implies the algebraic condition that G_1 and G_2 are conjugate, i.e., $S_1 \equiv S_2 \pmod{G}$ implies $G_1 \sim G_2 \pmod{G}$. Does the converse hold?

2.4.7. For any positive integer n, the group $\mathrm{Aut}(\widehat{\mathbf{C}})$ acts on the set of ordered n-tuples of distinct points of $\widehat{\mathbf{C}}$ by the rule $f(p_1, \ldots, p_n) = (f(p_1), \ldots, f(p_n))$. For what n is this action transitive? What are the orbits at the smallest n for which the action is intransitive? (This is a problem about cross-ratios in disguise, see for example [Jo-Si], Section 2.5.)

5. The Platonic solids and their rotations

We now work in $\mathbf{R}^3$.

At some point one encounters the Platonic solids—tetrahedron, cube, octahedron, dodecahedron, icosahedron—and hears that these are the only such. (See Figure 2.5.1.) This section constructs these five solids and shows that indeed they are exhaustive.

A Platonic solid is a regular convex polyhedron, meaning a compact intersection of finitely many half-spaces in $\mathbf{R}^3$ with congruent regular n-gon

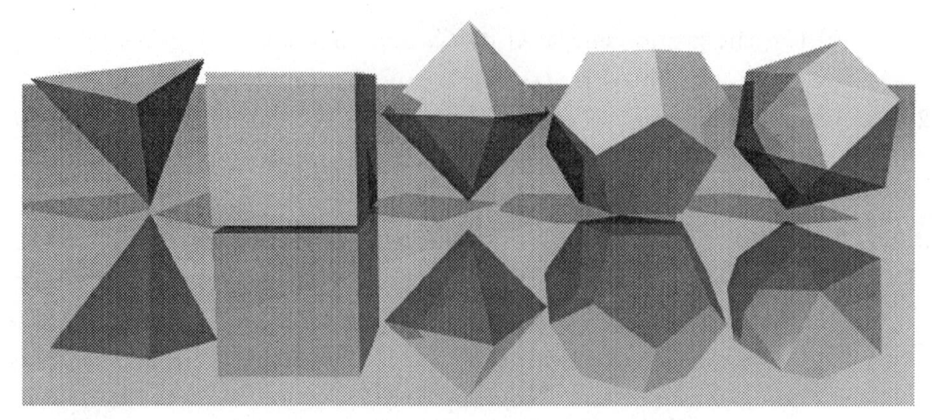

Figure 2.5.1. The Platonic solids

faces and some fixed number m of edges meeting at each vertex.

This definition is actually a bit naive. If the notion of polyhedron is generalized to allow nonconvex objects (see [Co 2], p.12), or to higher dimensions, the geometric and combinatorial conditions defining "regular" aren't so clear and natural. The right idea is that a polyhedron is regular if it has lots of symmetries, meaning rigid motions of $\mathbf{R}^3$ carrying the polyhedron back to itself. (See Exercise 2.2.6 for a description of the rigid motions of $\mathbf{R}^3$. In particular, placing a polyhedron with its center of gravity at the origin makes its symmetries orthogonal.) The symmetry group of a polyhedron, superficially extrinsic to the figure itself, gives the most natural description of its geometry. Incorporating these ideas into the definition makes it natural to study the symmetry groups of the Platonic solids in exhibiting the solids themselves.

(2.5.1) DEFINITION. *A* **flag** *of a polyhedron is a sequence* (v, e, f) *where* v *is a vertex,* e *is an edge containing* v*, and* f *is a face containing* e*. A* **Platonic solid** *is a polyhedron whose symmetry group acts transitively on its flags, that is, any flag can be taken to any other.*

This immediately forces a Platonic solid to have congruent regular n-gons for faces and some fixed number m of edges meeting at each vertex, as before (Exercise 2.5.2). The geometrical Exercise 2.5.3 limits the possible values of (m, n) to $(3,3)$, $(3,4)$, $(4,3)$, $(3,5)$, $(5,3)$. Since three equal angles fit together in at most one way at a vertex in space, there exists at most one Platonic solid, up to similarity, for any pair $(3, n)$. The **dual** of a Platonic solid, obtained by placing vertices at the face-centers of the original and connecting the pairs that came from adjacent faces, has the same symmetries as the original and its flags correspond bijectively with those of the original (Exercise 2.5.4), so it is another Platonic solid replacing (m, n) by (n, m). Thus Platonic solids with $n = 3$ are also unique. Every pair (m, n) above contains a 3, so each possible Platonic solid is unique. It remains to construct them.

The cube, with $(m, n) = (3, 4)$, is geometrically evident. Evident cube rotations are the identity, six rotations of order 4 (about each antipodal face-center pair), nine rotations of order 2 (about the three antipodal face-center pairs and the six antipodal mid-edge point pairs), and eight rotations of order 3 (two about each antipodal vertex pair). In fact, these are the only possible cube rotations. Indeed, any nonidentity cube rotation is about a point p, which lies in some face, nearest some vertex, as shown in Figure 2.5.2.

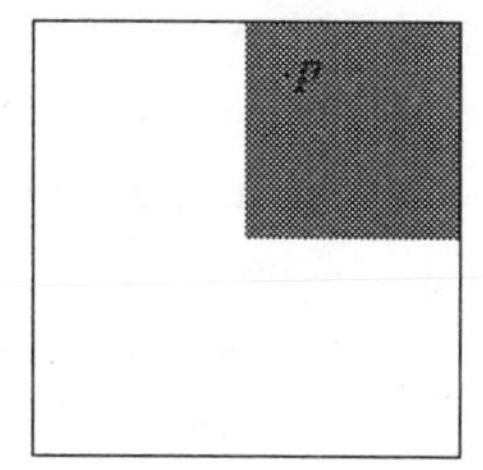

Figure 2.5.2. Center of rotation?

If p is a vertex, we have already counted the rotation. Otherwise the rotation permutes the vertices nearest p, moving them all (only p and its antipode are fixed), so p has at least two nearest vertices. This forces it to be a face-center or a mid-edge point, again giving a rotation already listed.

Thus the rotation group of the cube has 24 elements. The group is structurally a subgroup of S_4 by Exercise 2.4.4, because it acts faithfully on the set of four diagonals connecting antipodal cube vertices, so since $|S_4| = 24$, the cube rotations are all of S_4. The reflection R of $\mathbf{R}^3$ through the plane $x_1 = x_2$ is also a symmetry of the cube, so in fact the cube has 48 symmetries (see Exercise 2.5.5). Cube symmetries act on the set of cube flags and any flag has trivial stabilizer, so transitivity of the action now follows easily by counting that there are 48 flags (Exercise 2.5.6).

The octahedron, with $(m, n) = (4, 3)$, comes from dualizing the cube.

For the tetrahedron, with $(m, n) = (3, 3)$, connect a set of four non-neighboring vertices of the cube, as in Figure 2.5.3. Of the cube's rotations, the identity, three rotations of order 2 (about antipodal face-centers), and the eight rotations of order 3 preserve the tetrahedron, so the tetrahedral rotation group is A_4, the unique subgroup of index 2 in S_4. One can also

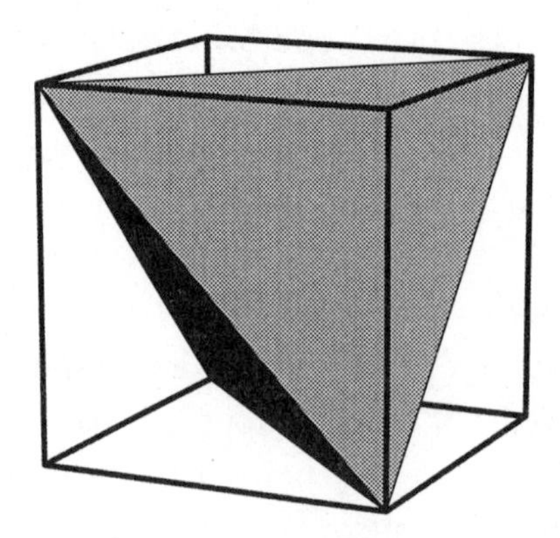

Figure 2.5.3. Tetrahedron

verify directly that the tetrahedral rotations act as A_4 on the cube diagonals. The reflection R is also a tetrahedral symmetry, so the full group of tetrahedral symmetries has order 24, and again transitivity is easy to check (Exercise 2.5.6).

Constructing the icosahedron, corresponding to $(m,n) = (5,3)$, takes more work. Let g denote the golden ratio, i.e., the positive root of $g^2 = 1-g$. (This g lies between 0 and 1, and is the reciprocal of what some authors call the golden ratio.) Center rectangles of sides 2, $2g$ in the three coordinate planes. Call the configuration $\mathcal{G}$. (See Figure 2.5.4.)

Consider the vertex $v = (1, -g, 0)$ in the figure. Its distance from any of

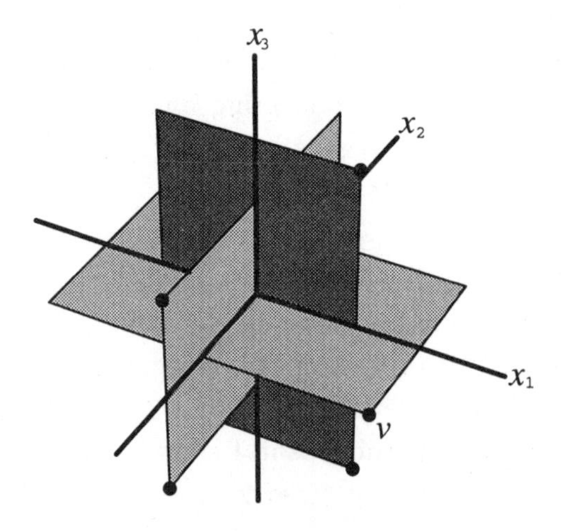

Figure 2.5.4. Golden configuration

$\{(0, -1, \pm g), (g, 0, \pm 1)\}$ (the four unlabeled dots in the figure) satisfies

$$d^2 = (g-1)^2 + g^2 + 1 = 2(g^2 + 1 - g) = 4g^2,$$

so $d = 2g$, which is also the distance from v to $(1, g, 0)$. All other vertices are farther from v. Thus five vertices are nearest to v; call such vertices **neighbors** of v. The rotations of $\mathcal{G}$ include those of order 2 about each axis and an order 3 rotation about $(1, 1, 1)$ (from the cube group), so they include the tetrahedral group A_4. The only rotation of $\mathcal{G}$ that stabilizes v is the identity, so since $\mathcal{G}$ has twelve vertices, the Orbit-Stabilizer Theorem says that its rotations are precisely A_4, which acts transitively on the vertices. It follows that each vertex has five neighbors at distance $2g$. Connecting neighbors gives the icosahedron (Figure 2.5.5), which will be a Platonic solid once we verify that its symmetries act transitively on its flags.

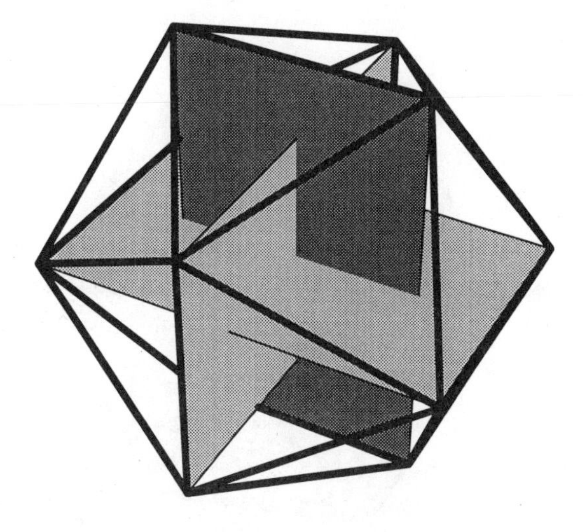

Figure 2.5.5. Icosahedron

Possible icosahedral rotations are limited to the identity, fifteen rotations of order 2 (about antipodal mid-edge point pairs), twenty rotations of order 3 (two about each antipodal face-center pair), and twenty-four rotations of order 5 (four about each antipodal vertex pair). The argument that these are the only possible rotations is similar to the case of the cube and is Exercise 2.5.9.

To show that the sixty rotations actually exist, return to the configuration $\mathcal{G}$ and the vertex v in Figure 2.5.4. The rotations of $\mathcal{G}$, which form a tetrahedral group A_4 as already shown, are certainly icosahedral rotations. A cyclic group C_5 of rotations about v by multiples of $2\pi/5$ is evident once we settle the slightly sticky point that the five neighbors of v are evenly distributed about a circle normal to the vector v. This holds because the icosahedral vertices lie on a sphere. Now, since A_4 and C_5 intersect trivially (their orders are relatively prime, and the order of a group element divides the order of the group), the set $A_4 \cdot C_5$ has sixty elements. So all sixty icosahedral rotations exist.

Reflecting $\mathbf{R}^3$ through any coordinate plane is visibly an icosahedral symmetry also. Thus the icosahedral group has 120 elements, and now transitivity on flags follows from counting them (Exercise 2.5.10).

To show that the icosahedral rotation group is abstractly A_5, first note that C_5 takes $\mathcal{G}$ to itself and four congruent golden configurations, each containing six icosahedral edges and the twelve icosahedral vertices. (See Figure 2.5.6.) These are the only possible such golden configurations. Call them $\mathcal{G}_1, \ldots, \mathcal{G}_5$.

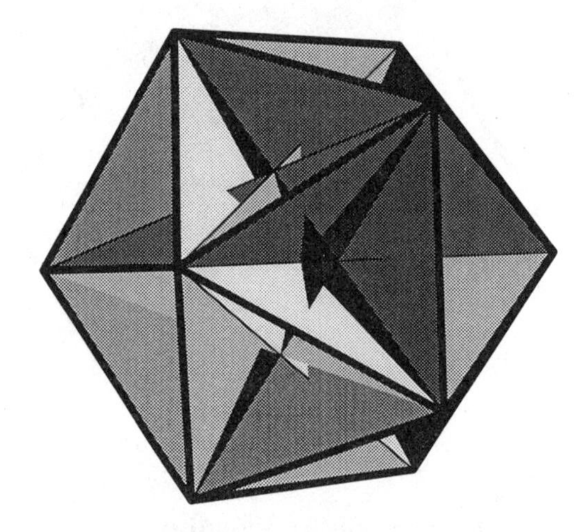

Figure 2.5.6. Five golden configurations

The icosahedral symmetries act faithfully on the five-element set $\{\mathcal{G}_i\}$. To see this, note that each icosahedral edge is a rectangular edge from some $\mathcal{G}_i$. Up to reindexing, the arrangement about any vertex is shown in Figure 2.5.7. Inspecting the figure shows that rotating the icosahedron about a face-center, a mid-edge point, or a vertex permutes the labels 1 through 5 nontrivially, and the permutation is even (Exercise 2.5.11). Thus the sixty icosahedral rotations are sixty even permutations in S_5, which must be A_5.

For more icosahedral geometry, Exercise 2.5.13 shows that the face-centers are the vertices of five intertwined tetrahedra. Finally, the dodeca-

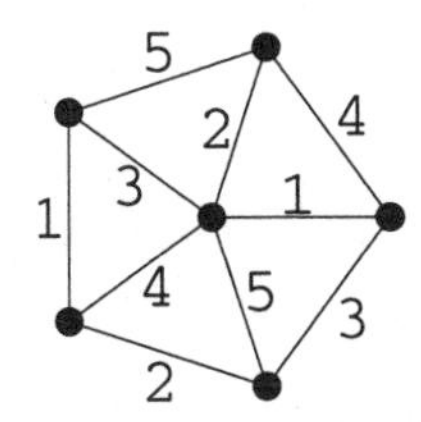

Figure 2.5.7. Edges of the five configurations

hedron, with $(m, n) = (3, 5)$, is the dual of the icosahedron.

Exercises

2.5.1. (a) Give a convex polyhedron with congruent faces that are not regular.

(b) Give a convex polyhedron with congruent regular n-gons for faces but not the same number of edges meeting at each vertex.

(c) Give a polyhedron with congruent regular n-gons for faces and m edges meeting at each vertex that is not a Platonic solid.

2.5.2. Show that a Platonic solid as defined in the text has congruent regular n-gons for faces and the same number of edges per vertex.

2.5.3. This exercise classifies the possible Platonic solids combinatorially.

(a) Show that the interior angles of a convex plane n-gon, not necessarily regular, sum to $(n-2)\pi$. (Sum the complementary external angles first.) In particular, the interior angle of a regular plane n-gon is $(1 - 2/n)\pi$.

(b) Illustrate this proof from Euclid that when three rays make a solid angle in space, any two of the angles between them exceed the third. Consider the configuration in Figure 2.5.8. The goal is to show that $\alpha + \beta > \gamma$. If $\alpha > \gamma$ there is nothing to show, so take $\alpha < \gamma$. Put a point E in the ABC-plane inside $\angle BAC$ so that $\angle BAE = \alpha$ and $\overline{AE} = \overline{AD}$. Adjusting C, we may assume B, E, C are collinear. By side-angle-side, triangles ABD and ABE are congruent, showing $BD = BE$. Also, $BD + DC > BC = BE + EC = BD + EC$, so $DC > EC$. Thus triangles ACD and ACE share the side AC, have equal sides AD and AE, while $DC > EC$. This shows that $\beta > \gamma - \alpha$ and the proof is complete.

(c) Now suppose m rays meet in convex fashion at a point v in space, where $m \geq 3$. Show that the angles at v sum to less than 2π by summing the angles α, β, θ over all triangles in Figure 2.5.9.

(d) If a Platonic solid has n-sided faces meeting m per vertex with

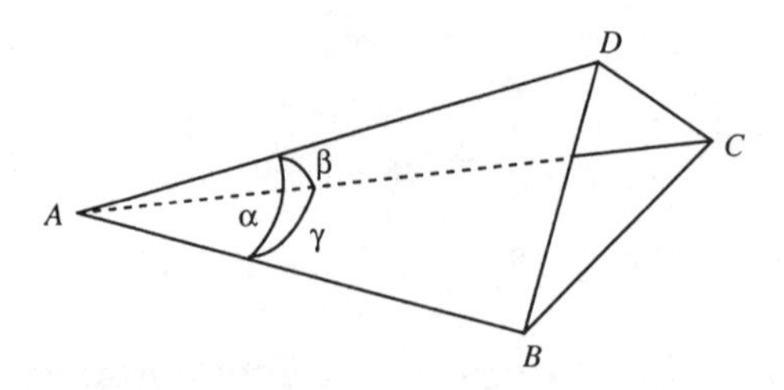

Figure 2.5.8. Solid angle

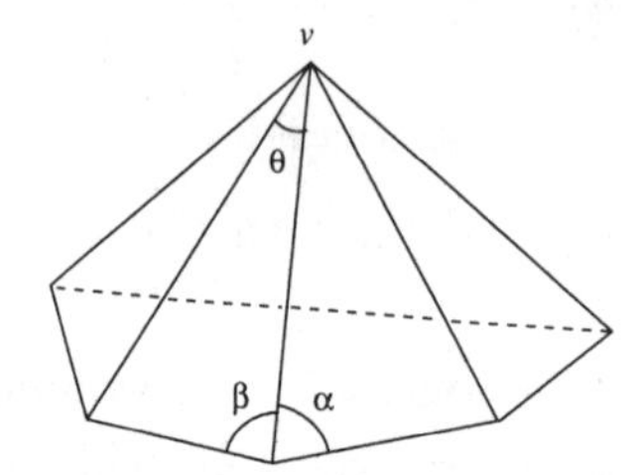

Figure 2.5.9. Angles at a vertex

$n, m \geq 3$, show that $m(1-2/n)\pi < 2\pi$ and consequently $(m-2)(n-2) < 4$. List the possibilities for (m, n).

2.5.4. Show that a Platonic solid and its dual have the same symmetries, and the flags of the dual correspond bijectively to the flags of the original.

2.5.5. A rigid motion of $\mathbf{R}^3$ that fixes the origin may be viewed as a matrix $A \in \mathrm{O}_3(\mathbf{R})$. The motion is a rotation if in fact $A \in \mathrm{SO}_3(\mathbf{R})$. Let G be a subgroup of $\mathrm{O}_3(\mathbf{R})$ and let $H = G \cap \mathrm{SO}_3(\mathbf{R})$, a subgroup of G. Show that $[G : H] = 1$ or 2 by considering the homomorphism $\det : G \longrightarrow \{\pm 1\}$. In particular, a Platonic solid has either as many or twice as many symmetries as it has rotations.

2.5.6. Show that the symmetries of the cube act transitively on flags. Do the same for the tetrahedron.

2.5.7. Write an explicit correspondence between the rotations of the cube and S_4. Check that the tetrahedral rotations map to A_4.

2.5.8. Show that the icosahedron has fifteen pairs of antipodal mid-edge points and ten pairs of antipodal face-centers.

2.5.9. Show that the only possible icosahedral rotations are those enumerated in the section.

2.5.10. Show that the icosahedral symmetries act transitively on flags.

2.5.11. Verify by inspecting Figure 2.5.7 that rotating the icosahedron about a face-center, a mid-edge point, or a vertex permutes the five golden configurations nontrivially, and the permutation is even.

2.5.12. A set of six antipodal pairs on a sphere with one common distance from every point to its five neighbors is an **icosahedral configuration**. This exercise shows that such a configuration is indeed the icosahedral vertices. Consider an icosahedral configuration $\mathcal{I}$ on the unit sphere. After a rotation the north pole $\mathbf{n}$ is in $\mathcal{I}$. Let p be one of its neighbors. The other four neighbors and p share a latitudinal circle S_1. Show that no point in an icosahedral configuration has antipodal neighbors, and use this to rule out the possibility that S_1 is the equator itself. The remaining six antipodal points are the south pole and five points on the opposite circle S_2. Show that no three points of S_1 are equidistant from p, and similarly for S_2, so p must have two neighbors on each circle. Argue similarly for the other neighbors of $\mathbf{n}$ to obtain a polyhedron with congruent equilateral triangular faces and five edges meeting at each vertex. Since $\mathcal{I}$ was completely specified by a pair of neighbors, any rotation of the sphere taking two neighbors back to points in $\mathcal{I}$ is in fact a polyhedral symmetry. These include all rotations by π about mid-edge points, by $\pm 2\pi/3$ about face-centers, and by $\pm 2\pi/5$ and $\pm 4\pi/5$ about vertices. The previous exercise shows transitivity on flags.

2.5.13. This exercise shows that the icosahedral face-centers are the vertices of five intertwined tetrahedra (see Figure 2.5.10). Recall from Exercise 2.4.5 that if the group G acts on the set S, $H \subset G$ is a

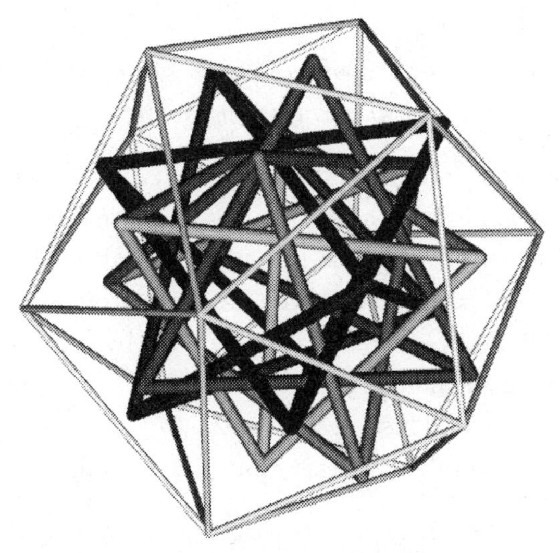

Figure 2.5.10. Five tetrahedra

subgroup, and $\mathcal{O}$ is an H-orbit containing the point s, then for any $\gamma \in G$, $\gamma\mathcal{O} \cap \mathcal{O} \neq \emptyset$ if and only if $\gamma \in H \cdot \text{stab}(s) \cdot H$. Working in the coordinate system of the golden rectangle configuration of Figure 2.5.4, specialize to G the group of icosahedral rotations, S the set of icosahedral face-centers, H the subgroup of 180-degree rotations about the coordinate axes, $\mathcal{O}$ the orbit of the face-center $s = (1,1,1)$ (after scaling). Show that $\mathcal{O}$ is the vertices of a tetrahedron. Show that stab(s) is cyclic of order 3 and $H \cdot \text{stab}(s) \cdot H$ is a tetrahedral group A_4. Let γ denote rotation about v by $2\pi/5$, of order 5. Show that $\gamma \notin H \cdot \text{stab}(s) \cdot H$, and therefore the sets $\mathcal{O}$, $\gamma\mathcal{O}$, ... , $\gamma^4\mathcal{O}$ are disjoint. Thus the twenty icosahedral face-centers form five tetrahedra. (Note: replacing s by the face-center $(1,1,-1)$ gives five different tetrahedra.)

6. Finite rotation groups of the sphere

We continue to work in $\mathbf{R}^3$. Along with some obvious groups, the Platonic rotations of the preceding section exhaust the possible finite rotation groups of the sphere.

(2.6.1) THEOREM. *Each finite rotation group of the sphere is isomorphic to one of the following groups:*

1. *Cyclic groups* $C_n = \langle s : s^n = 1\rangle$ *for* $n \geq 1$,
2. *Dihedral groups* $D_n = \langle s,t : s^n = t^2 = 1, tst = s^{-1}\rangle$ *for* $n \geq 2$,
3. *Platonic rotation groups* A_4, S_4 *and* A_5.

Any two finite rotation groups of the same form are conjugate, so these isomorphism classes are also the conjugacy classes.

PROOF. Let G be a finite rotation group. Take $|G| > 1$ since the case $|G| = 1$ is clear. The group G acts on the set S^2. By Exercise 2.6.1, G is cyclic if and only if there exists an orbit $\{p\}$ of length 1, and all cyclic rotation groups of the same order are conjugate. If the smallest orbit is $\{p, \tilde{p}\}$ of length 2, then in fact $\tilde{p} = -p$ and G is dihedral, generated by a rotation about p and a 180-degree rotation exchanging p and $-p$, and all such groups of the same order are also conjugate. This is Exercise 2.6.2.

To address the general case, start from the formula

$$|G| = 1 + \frac{1}{2}\sum_{p\in S^2}(|\text{stab}(p)| - 1) = 1 + \frac{1}{2}\sum_{p\in S^2}\left(\frac{|G|}{|\mathcal{O}_p|} - 1\right).$$

The first sum counts $G \setminus \{1_G\}$ since each nonidentity element of G stabilizes an antipodal pair in S^2. Nonzero summands come from points with degenerate orbits, meaning points p such that $|\mathcal{O}_p| < |G|$. Summing over such orbits rather than points gives

$$|G| = 1 + \frac{1}{2} \sum_{\mathcal{O}:|\mathcal{O}|<|G|} |\mathcal{O}| \left(\frac{|G|}{|\mathcal{O}|} - 1 \right), \tag{2.6.2}$$

so by some algebra, letting $n_{\mathcal{O}} = |G|/|\mathcal{O}| = |\mathrm{stab}(p)|$ for any $p \in \mathcal{O}$ (Exercise 2.6.3),

$$2\left(1 - \frac{1}{|G|}\right) = \sum_{\mathcal{O}:|\mathcal{O}|<|G|} \left(1 - \frac{1}{n_{\mathcal{O}}}\right). \tag{2.6.3}$$

Another flurry of algebra on the condition $2 \leq n_{\mathcal{O}} \leq |G|$ (which holds for each degenerate orbit) gives

$$\frac{k}{2} \leq \sum_{\mathcal{O}:|\mathcal{O}|<|G|} \left(1 - \frac{1}{n_{\mathcal{O}}}\right) \leq k\left(1 - \frac{1}{|G|}\right), \tag{2.6.4}$$

where k is the number of degenerate orbits; with (2.6.3) this gives $k = 2$ or $k = 3$ (Exercise 2.6.4).

All that's left to do is some arithmetic mopping up. If $k = 2$, (2.6.3) becomes $\frac{2}{|G|} = \frac{1}{n_1} + \frac{1}{n_2}$, forcing $n_1 = n_2 = |G|$ which implies two length-1 orbits, so G is cyclic. If $k = 3$, (2.6.3) becomes $1 + \frac{2}{|G|} = \frac{1}{n_1} + \frac{1}{n_2} + \frac{1}{n_3}$; indexing so that $n_1 \geq n_2 \geq n_3$ forces $n_3 = 2$ and the following possibilities for n_1, n_2 and $|G|$ (Exercise 2.6.5).

(2.6.5)

	n_1	n_2	n_3	$\lvert G\rvert$	$\lvert\mathcal{O}_1\rvert$	$\lvert\mathcal{O}_2\rvert$	$\lvert\mathcal{O}_3\rvert$
(a)	n	2	2	$2n$	2	n	n
(b)	3	3	2	12	4	4	6
(c)	4	3	2	24	6	8	12
(d)	5	3	2	60	12	20	30

In case (a), there is an orbit of order 2 and so G is dihedral. The other three cases are the rotation groups of the tetrahedron, octahedron, and icosahedron, with $\mathcal{O}_1$ the vertices, $\mathcal{O}_2$ the face-centers and $\mathcal{O}_3$ the mid-edges. For example, in case (d), let p be a point from the orbit $\mathcal{O}_1$ of length 12. Then since $\mathrm{stab}(-p) = \mathrm{stab}(p)$, $-p$ also sits in an orbit of length 12, which must be $\mathcal{O}_1$. Thus $\mathcal{O}_1$ consists of six antipodal pairs. Since $\mathrm{stab}(p)$ is cyclic of order 5, five points of $\mathcal{O}_1 \setminus \{p, -p\}$ are distributed evenly about a circle S_1 equidistant from p, and the other five points are their antipodes on the circle S_2 the same distance from $-p$, so p has five or ten neighbors in $\mathcal{O}_1$. The case of ten occurs only when $S_1 = S_2$ is the equator, but then each point $q \in \mathcal{O}_1 \setminus \{p, -p\}$ has only two neighbors, contradicting the number of neighbors of the generic point p. Thus all p have five neighbors, making $\mathcal{O}_1$ the vertices of an icosahedron (see Exercise 2.5.12) and the elements of G icosahedral rotations. Since $|G| = 60$, G is the full group A_5. Since all unit icosahedral configurations are congruent under rotation, all G of this form are conjugate by Exercise 2.4.6.

Cases (b) and (c) are similar. □

Since the cube and dodecahedron are the duals of the octahedron and icosahedron, the three Platonic groups in the theorem are exhaustive. No further reference will be made to the cube and dodecahedron.

Exercises

2.6.1. (a) Let G be a finite rotation group of S^2. Show that if G is cyclic then S^2 contains an orbit of length 1. Conversely, show that if S^2 contains an orbit $\{p\}$ of length 1, i.e., a point fixed by G, then G is a cyclic group of rotations about p. (Every nonidentity $g \in G$ takes the form $r_{p,\alpha}$ with $\alpha \in (0, 2\pi)$. Take the rotation s of minimal α and show it generates all of G.)
(b) Suppose that a cyclic group G as in part (a) has order n. Show that G is the rotation group of the set S containing p and n points distributed evenly about the equator between p and $-p$. Now use Exercise 2.4.6 to show that any two cyclic groups of the same order are conjugate.

2.6.2. (a) Let G be a finite rotation group of S^2 and suppose the smallest orbit is $\{p, \tilde{p}\}$ of length 2. Show that G is dihedral. (Let $H = \operatorname{stab}(p)$. Then $|H| = |G|/2$ and H is cyclic, generated by some s. If $|H| = 1$ then G is cyclic and S^2 has a length-1 orbit, contradiction. So $|H| > 1$ and the only other point H stabilizes is $-p$, showing $\tilde{p} = -p$. Any $t \in G \setminus H$ exchanges p and $-p$, so it is a 180-degree rotation about some point q on the equator between p and $-p$. Certainly s and t generate G. The product $stst$ fixes p and q, so it is the identity and G is dihedral.)
(b) As in the previous exercise, show any two dihedral groups of the same order are conjugate.

2.6.3. Obtain formula (2.6.3) from (2.6.2).

2.6.4. Explain the condition $2 \le n_{\mathcal{O}} \le |G|$, show how it implies (2.6.4), and show how (2.6.4) implies $2 \le k \le 3$.

2.6.5. Justify table (2.6.5).

2.6.6. Discuss cases (b) and (c) from table (2.6.5).

7. Projective representations of the finite rotation groups

For the sake of pending explicit calculations, this section exhibits conjugacy class representatives of the finite automorphism groups as subgroups of $\mathrm{PSU}_2(\mathbf{C})$. Doing so is a geometric process of putting the appropriate figures into standard position in $\mathbf{R}^3$ and then using Theorem 2.2.3 to write down a few rotation matrices.

For the cyclic group C_n, put the two one-point orbits at $\mathbf{n}$ and $\mathbf{s}$. Let $\zeta_m = e^{2\pi i/m}$ for any positive integer m. The generator is then

$$s_n = f_{\infty,2\pi/n} = \begin{bmatrix} \zeta_{2n} & 0 \\ 0 & \zeta_{2n}^{-1} \end{bmatrix} : z \mapsto \zeta_n z.$$

Each noncyclic group Γ is generated by two elements s_Γ and t_Γ. For the dihedral group D_n, put the two-point orbit at $\{\mathbf{n}, \mathbf{s}\}$ and put a fixed point of an order-2 element at $(1,0,0)$. Take s_n as in the cyclic group and

$$t_D = f_{1,\pi} = \begin{bmatrix} 0 & i \\ i & 0 \end{bmatrix} : z \mapsto \frac{1}{z}.$$

Then $D_n = \langle s_n, t_D \rangle$.

For the tetrahedral and octahedral groups, place a cube with its vertices on the sphere at $\{(\pm 1, \pm 1, \pm 1)/\sqrt{3}\}$. Number its diagonals 1 through 4 so that diagonal 1 passes through $(1,1,1)/\sqrt{3}$ and the indices increase counterclockwise about the four upper vertices. Take the tetrahedral vertices at $\{(\pm 1, \pm 1, \pm 1)/\sqrt{3} : x_1 x_2 x_3 > 0\}$. To generate the tetrahedral group Γ_T, set

$$s_T = f_{\infty,\pi} = \begin{bmatrix} i & 0 \\ 0 & -i \end{bmatrix} : z \mapsto -z,$$

which acts as $(13)(24)$ on the cube diagonals; set

$$t_T = f_{\pi((1,1,1)/\sqrt{3}),2\pi/3} = \frac{1}{2}\begin{bmatrix} 1+i & -1+i \\ 1+i & 1-i \end{bmatrix} : z \mapsto \frac{z+i}{z-i},$$

which acts as (243). Identifying rotations with the permutations they induce, compute (remembering to compose right to left)

$$t^{-1} s t = (234)(13)(24)(243) = (14)(23).$$

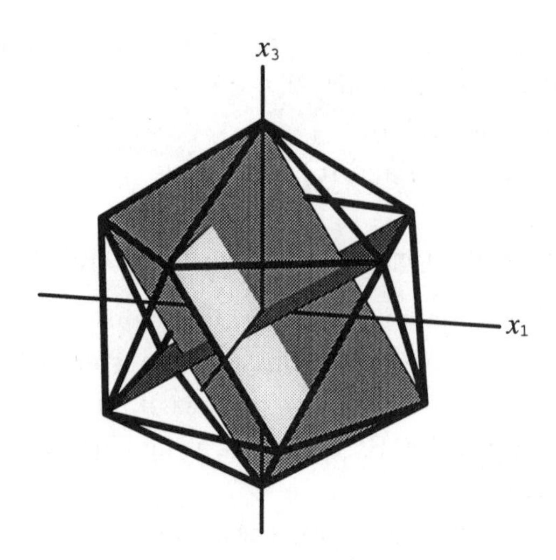

Figure 2.7.1. Rotated golden configuration

So $\langle s_T, t_T \rangle$ contains the **Klein group** $V = \{id, (12)(34), (13)(24), (14)(23)\}$ (also known as the **four-group**) and C_3, giving twelve elements and hence all of A_4.

To generate the octahedral group Γ_O, set

$$s_O = f_{\infty,\pi/2} = \frac{1}{\sqrt{2}} \begin{bmatrix} 1+i & 0 \\ 0 & 1-i \end{bmatrix} : z \mapsto iz,$$

which acts as (1234) on the cube diagonals; take $t_O = t_T$ from the tetrahedral group. Since $s_O^2 = s_T$, the group $\langle s_O, t_O \rangle$ contains the tetrahedral group A_4. Since s_O acts as an odd permutation, the group is all of S_4.

For the icosahedral group Γ_I, first rotate the golden rectangle configuration $\mathcal{G}$ by $r_{(0,-1,0),\arctan g}$ as shown in Figure 2.7.1. This puts two vertices at the poles and the other ten equally spaced along two latitudinal circles, cf. the proof of Theorem 2.6.1. With the vertices so positioned, the generators are a one-fifth rotation about the north pole,

$$s_I = f_{\infty,2\pi/5} = \begin{bmatrix} \zeta_{10} & 0 \\ 0 & \zeta_{10}^{-1} \end{bmatrix} : z \mapsto \zeta_5 z,$$

which acts as (12345) on the five golden configurations; and a 180-degree rotation about the mid-edge point $(-g, 0, 1)/|(-g, 0, 1)|$ (see Figure 2.7.2),

$$t_I = f_{\pi((-g,0,1)/|(-g,0,1)|),\pi} = \frac{-i}{\sqrt{2-g}} \begin{bmatrix} -1 & g \\ g & 1 \end{bmatrix} : z \mapsto \frac{g-z}{gz+1},$$

which acts as (23)(45). These generate the full group because ts acts as (135), and $(s^{-2}ts)t(s^{-1}ts^2)$ acts as (24)(35) (if the conjugation seems a little

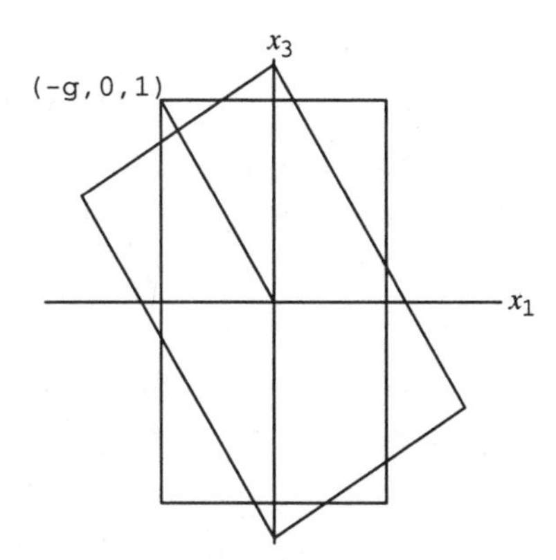

Figure 2.7.2. Rotated golden rectangle

mysterious, see Exercise 2.7.7 for its geometric genesis); thus $\langle s_I, t_I \rangle$ contains a Klein four-group and cyclic subgroups of orders 3 and 5, so its order is divisible by 60 and it is all of A_5.

The radial projections to S^2 of the tetrahedron, octahedron and icosahedron conveniently describe the degenerate orbits of the Platonic rotation groups (recall from the proof of Theorem 2.6.1 that an orbit is called dengenerate if it contains fewer elements than the group): these are the vertices, mid-edge points and face-centers. The degenerate orbits of the dihedral group D_n can be described analogously. As above, place the two-point orbit at $\{\mathbf{n}, \mathbf{s}\}$ and an n-point orbit equally spaced about the equator, including $(1,0,0)$. Connect $\mathbf{n}$ and $\mathbf{s}$ with great half-circles passing through the n-point orbit. The resulting **digon** has 2 vertices, n edges and n lunar faces, and the degenerate orbits of D_n are its vertices, mid-edge points and face-centers. Finally, the degenerate orbits of the cyclic group C_n, $\mathbf{n}$ and $\mathbf{s}$, occur at the vertex and "face-center" of the degenerate figure $\mathbf{n}$.

Exercises

2.7.1. Verify that $s_n^n = t_D^2 = id$ (the identity mapping) and $t_D s_n t_D = s_n^{-1}$.

2.7.2. Use Theorem 2.2.3 to confirm the various s and t transformations in this section.

2.7.3. Illustrate the spherical rotations that give rise to s_T, t_T and s_O, and confirm their actions on the four cube diagonals.

2.7.4. In the tetrahedral group, the map $t_T = f_{\pi((1,1,1)/\sqrt{3}),2\pi/3} : z \longrightarrow (z+i)/(z-i)$ maps the upper half plane $\mathcal{H}$ to $\mathbf{C} \setminus \mathcal{D}$, the exterior of the unit disk, since the upper half plane is precisely the points that are closer to i than to $-i$. Illustrate what's going on in terms of how

$r_{(1,1,1)/\sqrt{3},2\pi/3}$ rotates S^2.

2.7.5. Choose elements of A_4 other than (13)(24) and (243), e.g., (123), and find the corresponding elements of Γ_T in two ways: (1) by geometric inspection of how to rotate the tetrahedron, and (2) by multiplying powers of (13)(24) and (243) and then carrying out the corresponding multiplications of s_T and t_T.

2.7.6. Same as the preceding exercise but with S_4 and Γ_O.

2.7.7. Figure 2.7.3 shows the stereographic projections of the icosahedral vertices as repositioned in this section, cf. the cover of the book. The generators s_I, t_I for Γ_I are respectively a one-fifth counterclockwise revolution about the center-point, and a (non-Euclidean) half-revolution about the "1." Complete the labelling of the edges of the five golden configurations. Describe the geometric effect of s_I and t_I on the central pentagon and its edges. Confirm the actions of s_I and t_I on the five configurations. Explain geometrically why the conjugation $(s^{-2}ts)t(s^{-1}ts^2)$ acts as (24)(35).

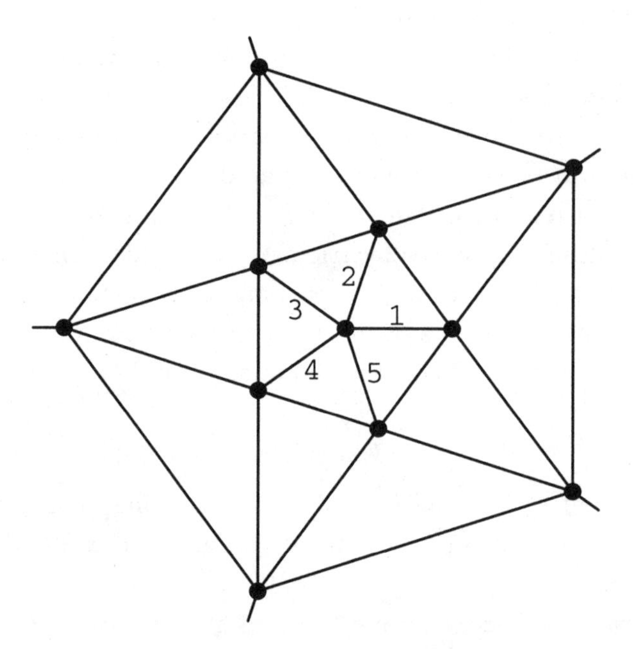

Figure 2.7.3. Projected icosahedron

2.7.8. List the (extended) complex values of the icosahedral vertices in the previous problem. They are $\{\infty, s_I^j t_I \infty, 0, s_I^j t_I 0 : 0 \leq j \leq 4\}$.

8. Summary

The automorphisms of the sphere are fractional linear transformations, best viewed as projective special matrix classes. In particular, Theorem 2.2.3 describes rotations in these terms. Any finite automorphism group of the sphere is conjugate to a rotation group, and geometry shows that the rotation group classes are one copy each of C_n, D_n, A_4, S_4, and A_5. These have convenient representations as projective matrix groups.

CHAPTER 3

Invariant functions

The conjugacy classes of finite automorphism groups now have normalized rotation group representatives, each preserving an associated geometric figure. By analogy, this chapter associates an algebraic object to each group. The idea is that like a geometric figure, a function f on $\widehat{\mathbf{C}}$ also can remain invariant under a group Γ of automorphisms. In symbols, the function f is Γ-invariant if

$$f \circ \gamma = f \qquad \text{for all } \gamma \in \Gamma.$$

This just means that the value of f at a point p depends only on the point's orbit $\mathcal{O}_p$ under Γ. The set of Γ-invariant rational functions, denoted $\mathbf{C}(Z)^{\Gamma}$, forms a subfield of $\mathbf{C}(Z)$.

For each rotation group Γ, this chapter constructs a normalized function f_{Γ} that generates the Γ-invariant rational functions, meaning that every Γ-invariant rational function is a rational expression in f_{Γ}. In symbols,

$$\mathbf{C}(Z)^{\Gamma} = \mathbf{C}(f_{\Gamma}).$$

The generator f_{Γ} takes distinct values on distinct orbits, thus identifying the set of orbits with the Riemann sphere.

Since rational functions correspond naturally to pairs of forms, which are convenient to work with, much of the computation is done with forms. The first three sections of the chapter are general, then the calculation becomes specific to each group.

Recommended reading: This material is drawn closely from Chapter I.2 of Klein [Kl]. See also Dickson [Di]. Cox, Little, and O'Shea [Co-Li-O'S] includes lots of nice material on invariants.

1. Invariant forms

Let Γ be any finite rotation group of the sphere.

(3.1.1) DEFINITION. *A meromorphic function $f : \widehat{\mathbf{C}} \longrightarrow \widehat{\mathbf{C}}$ is* **Γ-invariant** *if*

$$f \circ \gamma = f \qquad \text{for all } \gamma \in \Gamma.$$

As explained above, Γ-invariance means that f is really a function of Γ-orbits. To verify this definition it suffices to check the condition $f \circ \gamma = f$ for a set of generators γ of Γ (Exercise 3.1.1). For example, if C_n is the cyclic group generated by $s_n = \left[\begin{array}{cc} \zeta_{2n} & 0 \\ 0 & \zeta_{2n}^{-1} \end{array}\right] : z \mapsto \zeta_n z$ then the function $f(z) = z^n$ is C_n-invariant because

$$(f \circ s_n)(z) = f(\zeta_n z) = (\zeta_n z)^n = z^n = f(z)$$

and checking invariance under the generator s_n is sufficient.

Recall that a meromorphic function $f : \widehat{\mathbf{C}} \longrightarrow \widehat{\mathbf{C}}$ may be viewed as an algebraic mapping $f[Z_1 : Z_2] = [G(Z_1, Z_2) : H(Z_1, Z_2)] : \mathbf{P}^1(\mathbf{C}) \longrightarrow \mathbf{P}^1(\mathbf{C})$, where G and H are same-degree forms (homogeneous polynomials) with no common zeros. For example, the algebraic representation of $f(z) = z^n$ is $f[Z_1 : Z_2] = [Z_1^n : Z_2^n]$. The next task is to translate Definition 3.1.1 into a condition on the component forms G and H of f.

The group $\Gamma \subset \mathrm{PSL}_2(\mathbf{C})$, viewed as degree 1 algebraic mappings of $\mathbf{P}^1(\mathbf{C})$,

$$\gamma = \overline{\begin{bmatrix} a & b \\ c & d \end{bmatrix}} : [Z_1 : Z_2] \mapsto [aZ_1 + bZ_2 : cZ_1 + dZ_2],$$

lifts to a group $\Gamma' \subset \mathrm{SL}_2(\mathbf{C})$, viewed naturally as linear mappings of $\mathbf{C}^2$,

$$\gamma' = \begin{bmatrix} a & b \\ c & d \end{bmatrix} : (Z_1, Z_2) \mapsto (aZ_1 + bZ_2, cZ_1 + dZ_2).$$

Each $\gamma \in \Gamma$ has two lifts $\pm\gamma' \in \Gamma'$. For example, the generator s_n of C_n, with algebraic representation $s_n[Z_1 : Z_2] = [\zeta_{2n} Z_1 : \zeta_{2n}^{-1} Z_2]$, lifts to the mappings $\pm s_n'(Z_1, Z_2) = \pm(\zeta_{2n} Z_1, \zeta_{2n}^{-1} Z_2)$. The calculations

$$s_n'(\mathbf{C}^*(Z_1, Z_2)) = \mathbf{C}^*(\zeta_{2n} Z_1, \zeta_{2n}^{-1} Z_2),$$
$$-s_n'(\mathbf{C}^*(Z_1, Z_2)) = \mathbf{C}^*(\zeta_{2n} Z_1, \zeta_{2n}^{-1} Z_2)$$

show that the actions of the lifts $\pm s_n'$ on $\mathbf{C}^2 \setminus \{\mathbf{0}\}$ descend to $\mathbf{P}^1(\mathbf{C})$, where they agree with the original s_n. The same assertion holds in general: both lifts $\pm\gamma'$ of any $\gamma \in \Gamma$ act on $\mathbf{P}^1(\mathbf{C})$ as γ does. (Exercise 3.1.2 asks you to confirm this by recalling the commutative diagrams from Section 2.1.) As remarked in Section 2.1, this is just a finicky way of saying that constant multiples pass through linear functions and are absorbed by projective classes, so we may be casual about when we discard them. The point is that (Exercise 3.1.3) for all algebraic mappings $f = [G : H]$ and all $\gamma \in \Gamma$, the composition $f \circ \gamma$ expressed in components is

$$f \circ \gamma = [G \circ \gamma' : H \circ \gamma'] \qquad \text{for either lift } \gamma' \text{ of } \gamma.$$

For example, with $f = [Z_1^n : Z_2^n]$ and $s_n = [\zeta_{2n} Z_1 : \zeta_{2n}^{-1} Z_2]$ as above, we already know that

$$(f \circ s_n)[Z_1 : Z_2] = [(\zeta_{2n} Z_1)^n : (\zeta_{2n}^{-1} Z_2)^n] = [-Z_1^n : -Z_2^n] = f[Z_1 : Z_2].$$

But note that componentwise, $f \circ s_n$ agrees with f only up to the factor -1. More generally (Exercise 3.1.3 continued),

(3.1.2) PROPOSITION. *The algebraic mapping* $f = [G : H] : \mathbf{P}^1(\mathbf{C}) \longrightarrow \mathbf{P}^1(\mathbf{C})$ *is* Γ*-invariant if and only if for each* $\gamma' \in \Gamma'$ *there exists some nonzero complex number* $\chi(\gamma')$ *such that*

$$G \circ \gamma' = \chi(\gamma') G \quad \text{and} \quad H \circ \gamma' = \chi(\gamma') H.$$

In other words, f *is* Γ*-invariant if and only if* G *and* H *both transform by some common factor* χ *under* Γ'*.*

Thus to study Γ-invariant algebraic mappings f we may study forms and their transformation under Γ'. We are interested in forms satisfying the following condition:

(3.1.3) DEFINITION. *Let* $F \in \mathbf{C}[Z_1 : Z_2]$ *be a nonzero form. Suppose that for some function* $\chi_F : \Gamma' \longrightarrow \mathbf{C}^*$,

$$F \circ \gamma' = \chi_F(\gamma') F \qquad \text{for all } \gamma' \in \Gamma'.$$

Then F *is called* **Γ'-invariant with character** χ_F.

Exercise 3.1.4 shows that for such F, the character χ_F is a homomorphism from Γ' to $\mathbf{C}^*$ and is therefore specified by its action on the generators of Γ'. The image of χ_F sits in the $|\Gamma'|$th roots of unity. If χ takes the same value on both lifts $\pm\gamma'$ of each $\gamma \in \Gamma$, then it descends to a well-defined character of Γ, also called χ_F. Since $-\gamma' = (-I)\gamma'$ (where I is the 2-by-2 identity matrix), and χ_F is a homomorphism, and the calculation

$$(F \circ (-I))(Z_1, Z_2) = F(-Z_1, -Z_2) = (-1)^{\deg(F)} F(Z_1, Z_2)$$

shows that $\chi_F(-I) = (-1)^{\deg(F)}$, the character χ_F descends to Γ exactly when $\deg(F)$ is even.

Returning to the C_n-invariant example $f = [Z_1^n : Z_2^n]$, the computation that $f \circ s_n = [-Z_1^n : -Z_2^n]$ shows that the component forms are C_n'-invariant with character $\chi : s'_n \mapsto -1$. The homomorphic properties of χ (or another direct computation—see Exercise 3.1.5) give $\chi : -s'_n \mapsto (-1)^{n+1}$, so indeed χ descends to C_n exactly when n is even.

The remainder of this section will establish a theorem that makes it easy to create and recognize Γ'-invariant forms. Since the second variable in a form is essentially redundant, the next result is merely a rephrase of the Fundamental Theorem of Algebra.

(3.1.4) PROPOSITION. *Any nonzero $F \in \mathbf{C}[Z_1 : Z_2]$ factors into linear forms that are unique up to constant multiples.*

PROOF. Let $d = \deg(F)$ and let e be the largest degree of Z_2 dividing F. Using the homogenization and dehomogenization operators from Section 1.4 (in particular, see Exercise 1.4.6) gives

$$\begin{aligned} F(Z_1, Z_2) &= Z_2^e(F_*(Z))^* && \text{with } \deg F_* = d - e \\ &= a_{d-e} Z_2^e \Big(\prod_{i=1}^{d-e}(Z - r_i)\Big)^* && \text{with } a_{d-e} \neq 0 \\ &= a_{d-e} Z_2^e \prod_{i=1}^{d-e}(Z_1 - r_i Z_2) = a_{d-e}\prod_{i=1}^{d}(s_i Z_1 - r_i Z_2) \end{aligned}$$

with $r_1, \ldots, r_{d-e}$ uniquely determined, $s_1, \ldots, s_{d-e} = 1$, $r_{d-e+1}, \ldots, r_d = -1$, and $s_{d-e+1}, \ldots, s_d = 0$. □

Each linear factor $s_i Z_1 - r_i Z_2$ of a form F vanishes on the projective class $\mathbf{C}^*(r_i, s_i)$, so the zeros of F in $\mathbf{C}^2 \setminus \{\mathbf{0}\}$—call them $\mathcal{Z}(F)$—are a finite **multiset** (meaning elements can repeat—see Exercise 3.1.6 for some formalism) in $\mathbf{P}^1(\mathbf{C})$; this multiset depends only on the **form-class** of F, $\mathrm{cl}(F) = \{\lambda F : \lambda \in \mathbf{C}^*\}$. Conversely, a finite multiset $\mathcal{Z}$ in $\mathbf{P}^1(\mathbf{C})$ specifies the form-class $\mathrm{cl}(\prod_{[r:s]\in\mathcal{Z}}(sZ_1 - rZ_2))$ rather than any particular form. Thus nonzero form-classes and finite multisets in $\mathbf{P}^1(\mathbf{C})$ are in bijective correspondence.

Since either lift γ' of any element $\gamma \in \Gamma$ acts on projective classes in $\mathbf{C}^2 \setminus \{\mathbf{0}\}$ as γ acts on $\mathbf{P}^1(\mathbf{C})$, it follows that for any form F, $\mathcal{Z}(F \circ \gamma') = \gamma^{-1}\mathcal{Z}(F)$: this is clear for constant forms; it holds for linear forms (Exercise 3.1.7), whose form-classes correspond to points in $\mathbf{P}^1(\mathbf{C})$; and it follows in general since products of form classes correspond to unions of multisets. Now we can state the theorem.

(3.1.5) THEOREM. *A nonzero form F is Γ'-invariant if and only if $\mathcal{Z}(F)$ is a multiunion of Γ-orbits in $\mathbf{P}^1(\mathbf{C})$.*

This follows from the fact that F is Γ'-invariant if and only if $\text{cl}(F\circ\gamma') = \text{cl}(F)$—so correspondingly $\mathcal{Z}(F\circ\gamma') = \mathcal{Z}(F)$—for all $\gamma' \in \Gamma'$, and from the relation $\mathcal{Z}(F\circ\gamma') = \gamma^{-1}\mathcal{Z}(F)$. Exercise 3.1.8 asks for the details.

Exercises

3.1.1. Show that a meromorphic function f is Γ-invariant if and only if it is invariant under a set of generators for Γ.

3.1.2. In the notation of the section, use the commutative diagrams (2.1.1) and (2.1.2) (specialized to determinant 1) to show that both lifts $\pm\gamma' \in \Gamma'$ of $\gamma \in \Gamma$ act on projective classes in $\mathbf{C}^2 \setminus \{\mathbf{0}\}$, i.e., on $\mathbf{P}^1(\mathbf{C})$, as γ. That is, show that $\pm\gamma'(\mathbf{C}^*(Z_1, Z_2)) = \gamma[Z_1 : Z_2]$ for all $[Z_1 : Z_2] \in \mathbf{P}^1(\mathbf{C})$.

3.1.3. Let $f = [G : H]$ be an algebraic mapping of $\mathbf{P}^1(\mathbf{C})$. Use the preceding exercise to show that $f\circ\gamma = [G\circ\gamma' : H\circ\gamma']$ for all $\gamma \in \Gamma$ and either lift γ' of γ, so that $f\circ\gamma = f$ if and only if $[G\circ\gamma' : H\circ\gamma'] = [G : H]$ for all $\gamma' \in \Gamma'$. Prove Proposition 3.1.2 by showing that this second condition is equivalent to G and H being Γ'-invariant with the same character. (This last equivalence is actually a bit subtle. Since G and H are relatively prime, the condition implies that $G\circ\gamma' = \chi(\gamma')G$ and $H\circ\gamma' = \chi(\gamma')H$ for some $\chi(\gamma') \in \mathbf{C}[Z_1 : Z_2]$. Comparing degrees shows that $\chi(\gamma')$ is constant.)

3.1.4. If the nonzero form F is Γ'-invariant, show that the character $\chi_F : \Gamma' \longrightarrow \mathbf{C}^*$ such that $F\circ\gamma' = \chi_F(\gamma')F$ for all $\gamma' \in \Gamma'$ is a homomorphism. Show that $\chi_F(-I) = (-1)^{\deg(F)}$, so the character χ_F descends to Γ if and only if $\deg(F)$ is even.

3.1.5. Let $f = [Z_1^n : Z_2^n]$, $s_n[Z_1 : Z_2] = [\zeta_{2n}Z_1 : \zeta_{2n}^{-1}Z_2]$, and $-s_n'(Z_1, Z_2) = (-\zeta_{2n}Z_1, -\zeta_{2n}^{-1}Z_2)$. Since f is C_n-invariant, its component forms must be C_n'-invariant with some character χ. The section showed that $\chi(s_n') = -1$. Since χ is a homomorphism, it follows that $\chi(-s_n') = (-1)^n(-1)$, but show this by direct computation.

3.1.6. Define a multiset in a set S to be a formal sum $\sum_{s\in S} n_s \cdot s$ (with each $n_s \in \mathbf{N}$), interpreted as n_s copies of each element s. What is the cardinality of the multiset $\sum n_s \cdot s$? What are the union and intersection of the multisets $\sum n_s \cdot s$ and $\sum m_s \cdot s$?

3.1.7. Show that for any linear form F, $\mathcal{Z}(F\circ\gamma') = \gamma^{-1}\mathcal{Z}(F)$. (Show that both sides are the same point in $\mathbf{P}^1(\mathbf{C})$—recall that γ' effectively acts on $\mathbf{P}^1(\mathbf{C})$ as γ does.)

3.1.8. Fill in the proof of Theorem 3.1.5.

2. Orbit-forms and invariant forms

For each finite rotation group Γ, a **full orbit-form** is a form of degree $|\Gamma|$ vanishing on exactly one orbit, with a zero of the same order at each orbit point. Such a form is Γ'-invariant by Theorem 3.1.5. To create such forms, lift each $\gamma \in \Gamma$ to a linear mapping $\gamma' = (\gamma'_1, \gamma'_2)$. Then for any point $p = [p_1 : p_2] \in \mathbf{P}^1(\mathbf{C})$, the full orbit-form vanishing on the orbit $\mathcal{O}_p$ is

$$F_p(Z_1, Z_2) = \prod_{\gamma \in \Gamma} (\gamma'_2(p_1, p_2)Z_1 - \gamma'_1(p_1, p_2)Z_2).$$

This form has repeated factors when p has nontrivial stabilizer in Γ, i.e., when $\mathcal{O}_p$ is degenerate, but in any case the degree of F_p is $|\Gamma|$. Thus the "full" in "full orbit-form" refers to form degree, not orbit length. The form F_p is really only defined up to form-class, but rather than maintain such a fussy distinction we work with class representatives from now on, freely discarding constant factors.

In the case of the group C_n whose generator s_n lifts to the mapping $s'_n(p_1, p_2) = (\zeta_{2n}p_1, \zeta_{2n}^{-1}p_2)$, the full orbit-form associated to the point $p = [p_1 : p_2]$ is

$$F_p(Z_1, Z_2) = \prod_{i=0}^{n-1} (\zeta_{2n}^{-i}p_2 Z_1 - \zeta_{2n}^{i}p_1 Z_2).$$

Since this vanishes precisely when $[Z_1 : Z_2]$ lies in $\mathcal{O}_p = \{\zeta_n^i p : 0 \leq i < n\}$, it multiplies out to a constant times $p_2^n Z_1^n - p_1^n Z_2^n$ (Exercise 3.2.1).

All full orbit-forms share the same character. To see this, note that the complex coefficients of $F_p \in \mathbf{C}_{|\Gamma|}[Z_1 : Z_2]$ vary continuously with $p \in \mathbf{P}^1(\mathbf{C})$. Thus for any $\gamma' \in \Gamma'$, the corresponding character value $\chi_p(\gamma') = (F_p \circ \gamma')/F_p$ also varies continuously with p. But all characters χ_p have images in the same discrete subset of $\mathbf{C}^*$ (the $|\Gamma'|$th roots of unity, as discussed in Section 3.1), so in fact $\chi_p(\gamma')$ is locally constant with respect to p. Since any two points in $\mathbf{P}^1(\mathbf{C})$ are joined by a path, $\chi_p(\gamma')$ is constant over all of $\mathbf{P}^1(\mathbf{C})$.

In the case of C_n, the calculations in the preceding section show that the forms $G(Z_1, Z_2) = Z_1^n$ and $H(Z_1, Z_2) = Z_2^n$ transform by the same character $\chi : s'_n \mapsto -1$; so does any linear combination of G and H. Therefore χ is indeed the character for all full orbit-forms $F_p = p_2^n G - p_1^n H$.

For the rest of this chapter, χ_Γ denotes the full orbit-form character for each finite rotation group Γ. This notation is slightly misleading since χ_Γ is actually defined on Γ', but χ_Γ descends to Γ whenever $|\Gamma|$ is even, which holds for all Γ except C_n, n odd.

Recall that each finite rotation group Γ has associated degenerate orbits $\mathcal{O}_1$, $\mathcal{O}_2$, and $\mathcal{O}_3$ (or just the first two for the cyclic groups) with each $|\mathcal{O}_i| < |\Gamma|$. While the full orbit-form for a nondegenerate orbit $\mathcal{O}$ is simply $F = \prod_{[p_1:p_2]\in\mathcal{O}}(p_2Z_1 - p_1Z_2)$, the full orbit-form for a degenerate orbit $\mathcal{O}_i$ is a power $F_i^{n_i}$ where

$$F_i = \prod_{[p_1:p_2]\in\mathcal{O}_i} (p_2Z_1 - p_1Z_2) \tag{3.2.1}$$

has degree $|\mathcal{O}_i|$, and $n_i = |\Gamma|/|\mathcal{O}_i|$. These lower degree forms F_i are called **degenerate** orbit-forms. Like full orbit-forms, they are invariant by Theorem 3.1.5. Their associated characters are denoted χ_i. Each $\chi_i^{n_i}$ is the full orbit-form character χ_Γ.

For the cyclic group C_n, the degenerate orbits are $\mathcal{O}_1 = \{\mathbf{s}\} = \{[0:1]\}$ and $\mathcal{O}_2 = \{\mathbf{n}\} = \{[1:0]\}$. The corresponding degenerate orbit-forms (3.2.1) are simply $F_1 = Z_1$ and $F_2 = Z_2$. To find their characters, compute

$$(F_1 \circ s_n')(Z_1, Z_2) = F_1(\zeta_{2n}Z_1, \zeta_{2n}^{-1}Z_2) = \zeta_{2n}Z_1 = \zeta_{2n}F_1(Z_1, Z_2),$$

so $\chi_1 : s_n' \mapsto \zeta_{2n}$. Similarly (Exercise 3.2.2) $\chi_2 : s_n' \mapsto \zeta_{2n}^{-1}$. These characters do not descend to C_n.

The degenerate orbit-forms are convenient because they generate all full orbit-forms—in fact, we only need two of them.

(3.2.2) THEOREM. *The full orbit-forms are the linear combinations*

$$F = \lambda_1 F_1^{n_1} + \lambda_2 F_2^{n_2}, \qquad [\lambda_1 : \lambda_2] \in \mathbf{P}^1(\mathbf{C}).$$

PROOF. A form $F = \lambda_1 F_1^{n_1} + \lambda_2 F_2^{n_2}$ is Γ'-invariant with character χ_Γ and degree $|\Gamma|$. Such F is visibly a full orbit-form if $\lambda_1 = 0$ or $\lambda_2 = 0$. Otherwise it vanishes on a multiunion of orbits of total length $|\Gamma|$, none of which is $\mathcal{O}_1$ or $\mathcal{O}_2$. Thus F is $F_3^{n_3}$ or the form of a single full orbit, giving a full orbit-form in either case.

Conversely, take the orbit $\mathcal{O}_p$ of any point $p = [p_1 : p_2]$, and set $[\lambda_1 : \lambda_2] = [F_2^{n_2}(p_1, p_2) : -F_1^{n_1}(p_1, p_2)]$. Then $F = \lambda_1 F_1^{n_1} + \lambda_2 F_2^{n_2}$ vanishes at p and, as just argued, must be the full orbit-form vanishing on $\mathcal{O}_p$. □

Since the invariant forms are the products of orbit-forms, the general invariant form is (Exercise 3.2.3)

$$F = F_1^{e_1} F_2^{e_2} F_3^{e_3} \prod_{\lambda\in\mathbf{P}^1(\mathbf{C})} (\lambda_1 F_1^{n_1} + \lambda_2 F_2^{n_2})^{e_\lambda} \tag{3.2.3}$$

with $0 \le e_i < n_i$ for $i = 1, 2, 3$ and $e_\lambda = 0$ for all but finitely many λ. (Set $F_3 = 1$ in the cyclic case to make this notation work.) The associated character is $\chi_F = \chi_1^{e_1}\chi_2^{e_2}\chi_3^{e_3}\chi_\Gamma^e$ where $e = \sum_\lambda e_\lambda$, and $\chi_3 = 1$ for the cyclic group.

Thus to express all invariant forms in a simple fashion, it suffices to find the degenerate orbit-forms F_1, F_2, and F_3. Some classical machinery, introduced in the next section, carries this out handily. Theorem 3.2.2 shows that the three forms $F_i^{n_i}$ for any noncyclic group are linearly dependent, so for each such group there is a linear relation (called a **syzygy**)

$$F_1^{n_1} - aF_2^{n_2} - bF_3^{n_3} = 0.$$

Finding this syzygy only requires matching terms containing the highest and next highest powers of Z_1.

Exercises

3.2.1. Let C_n be the cyclic group with generator s_n lifting to the map $s_n'(p_1, p_2) = (\zeta_{2n}p_1, \zeta_{2n}^{-1}p_2)$. Show that the full orbit-form F_p associated to the point $p = [p_1 : p_2]$ is a constant times $p_2^n Z_1^n - p_1^n Z_2^n$.

3.2.2. Compute the character χ_2 for the cyclic group C_n.

3.2.3. Prove that the general invariant form F is given by (3.2.3). (Hint: since F is a product of degenerate and full orbit-forms, Theorem 3.2.2 gives the result except for the conditions on e_i, $i = 1, 2, 3$; but also the forms $F_i^{n_i}$, $i = 1, 2, 3$ are full, so carrying out the division algorithm on the multiplicity of the degenerate orbit-forms in F completes the proof.)

3.2.4. Write the general invariant form for the cyclic group C_n.

3. Covariant forms

A **covariant** with respect to a group Γ' of linear maps on $\mathbf{C}^2$ is a function $C : \mathbf{C}[Z_1 : Z_2] \longrightarrow \mathbf{C}[Z_1 : Z_2]$ satisfying

$$C(F \circ \gamma') = (CF) \circ \gamma' \quad \text{for all } F \in \mathbf{C}[Z_1 : Z_2] \text{ and } \gamma' \in \Gamma'.$$

If for some exponent e, $C(aF) = a^e CF$ for all $a \in \mathbf{C}$ and all $F \in \mathbf{C}[Z_1 : Z_2]$, then for any Γ'-invariant F with character χ_F,

$$(CF) \circ \gamma' = C(F \circ \gamma') = C(\chi_F(\gamma')F) = \chi_F(\gamma')^e CF \quad \text{for all } \gamma' \in \Gamma',$$

showing that CF is also Γ'-invariant, with character χ_F^e.

The two covariants that we need are the **Hessian** and the **Jacobian**.

Take the matrix

$$M_H(F) = \begin{bmatrix} D_{11}F & D_{12}F \\ D_{21}F & D_{22}F \end{bmatrix},$$

where D_{11} means differentiating twice with respect to the first variable, etc. This is the Hessian matrix of F from multivariable calculus. The Hessian of a form F is $HF = \det(M_H(F))$. Since differentiating reduces form degree by 1, HF is a form of degree $2(\deg(F) - 2)$ (Exercise 3.3.2). To see that H is a covariant with respect to $\mathrm{SL}_2(\mathbf{C})$, take any linear mapping $\gamma' = (\gamma_1', \gamma_2')$ with matrix $[\gamma']$ of determinant 1. By two applications of the chain rule, the entries of $M_H(F)$ are

$$\begin{aligned}
[M_H(F \circ \gamma')]_{ij} &= D_{ij}(F \circ \gamma') = D_j \sum_{k=1}^{2} ((D_k F) \circ \gamma') D_i \gamma_k' \\
&= \sum_{k=1}^{2} (D_j((D_k F) \circ \gamma'))[\gamma']_{ki} = \sum_{k=1}^{2} \sum_{l=1}^{2} ((D_{kl}F) \circ \gamma') D_j \gamma_l' [\gamma']_{ki} \\
&= \sum_{k=1}^{2} \sum_{l=1}^{2} [\gamma']_{ik}^t ((D_{kl}F) \circ \gamma')[\gamma']_{lj} = ([\gamma']^t (M_H(F) \circ \gamma')[\gamma'])_{ij}
\end{aligned}$$

where $M_H(F) \circ \gamma' = \begin{bmatrix} (D_{11}F) \circ \gamma' & (D_{12}F) \circ \gamma' \\ (D_{21}F) \circ \gamma' & (D_{22}F) \circ \gamma' \end{bmatrix}$. This implies

$$H(F \circ \gamma') = \det([\gamma']^t (M_H(F) \circ \gamma')[\gamma']) = \det(M_H(F) \circ \gamma') = (HF) \circ \gamma',$$

so the Hessian is a covariant. Also, $H(aF) = a^2 HF$, so if F is Γ'-invariant with character χ_F, then HF is Γ'-invariant with character $\chi_{HF} = \chi_F^2$.

The Jacobian of a form F is the determinant

$$JF = \det \begin{bmatrix} D_1 F & D_2 F \\ D_1 HF & D_2 HF \end{bmatrix}.$$

The matrix is the Jacobian matrix for the vector-valued function (F, HF), again from multivariable calculus. The Jacobian of F is a form of degree $3(\deg(F) - 2)$ (Exercise 3.3.2 again). If γ' is as above, then similarly to the computation of the Hessian,

$$\begin{aligned}
J(F\circ\gamma') &= \det\begin{bmatrix} D_1(F\circ\gamma') & D_2(F\circ\gamma') \\ D_1(HF\circ\gamma') & D_2(HF\circ\gamma') \end{bmatrix} \\
&= \det\begin{bmatrix} (D_1F\circ\gamma')[\gamma']_{11}+ & (D_1F\circ\gamma')[\gamma']_{12}+ \\ (D_2F\circ\gamma')[\gamma']_{21} & (D_2F\circ\gamma')[\gamma']_{22} \\ & \\ (D_1HF\circ\gamma')[\gamma']_{11}+ & (D_1HF\circ\gamma')[\gamma']_{12}+ \\ (D_2HF\circ\gamma')[\gamma']_{21} & (D_2HF\circ\gamma')[\gamma']_{22} \end{bmatrix} \\
&= \det\left(\begin{bmatrix} (D_1F)\circ\gamma' & (D_2F)\circ\gamma' \\ (D_1HF)\circ\gamma' & (D_2HF)\circ\gamma' \end{bmatrix}[\gamma']\right) \\
&= \det\begin{bmatrix} (D_1F)\circ\gamma' & (D_2F)\circ\gamma' \\ (D_1HF)\circ\gamma' & (D_2HF)\circ\gamma' \end{bmatrix} \\
&= (JF)\circ\gamma',
\end{aligned}$$

so the Jacobian is also a covariant. Since $J(aF) = a^3JF$, if F is Γ'-invariant with character χ_F then JF is Γ'-invariant with character $\chi_{JF} = \chi_F^3$.

Exercises

3.3.1. Compute the Hessian and Jacobian of the forms $F = Z_1^n + Z_2^n$, $F = Z_1^nZ_2^m$.

3.3.2. Show that the Hessian and Jacobian are forms with $\deg(HF) = 2(\deg(F) - 2)$ and $\deg(JF) = 3(\deg(F) - 2)$.

3.3.3. Write a computer algebra routine to compute Hessians and Jacobians. (This will be quite useful in the next section.)

4. Calculation of the degenerate orbit-forms

As we have already seen, the degenerate orbits $\mathcal{O}_1 = \{\mathbf{s}\}$ and $\mathcal{O}_2 = \{\mathbf{n}\}$ for the cyclic group C_n have orbit-forms

$$F_{1,C_n} = Z_1,$$
$$F_{2,C_n} = Z_2.$$

Lifting the generator s_n of C_n to $s'_n : (Z_1, Z_2) \mapsto (\zeta_{2n}Z_1, \zeta_{2n}^{-1}Z_2)$ gave $F_{1,C_n} \circ s'_n = \zeta_{2n}F_{1,C_n}$, and similarly $F_{2,C_n} \circ s'_n = \zeta_{2n}^{-1}F_{2,C_n}$. Thus the characters for these forms and the general orbit-form character are as before,

$$\chi_{1,C_n} : s'_n \mapsto \zeta_{2n},$$
$$\chi_{2,C_n} = \chi_{1,C_n}^{-1},$$
$$\chi_{1,C_n}^n = \chi_{2,C_n}^n = \chi_{C_n} : s'_n \mapsto -1.$$

The first two characters do not descend to C_n; the last one descends when n is even.

The other groups have three degenerate orbits each: the vertex orbit $\mathcal{O}_1$, the face orbit $\mathcal{O}_2$, and the edge orbit $\mathcal{O}_3$. For the dihedral group, these are $\mathcal{O}_1 = \{\mathbf{n}, \mathbf{s}\}$, $\mathcal{O}_2 = \{\zeta_n^j \zeta_{2n} : 0 \leq j < n\}$ and $\mathcal{O}_3 = \{\zeta_n^j : 0 \leq j < n\}$, with orbit-forms (3.2.1) given by

$$F_{1,D_n} = Z_1 Z_2,$$
$$F_{2,D_n} = \prod_{j=0}^{n-1} (Z_1 - \zeta_n^j \zeta_{2n} Z_2) = Z_1^n + Z_2^n,$$
$$F_{3,D_n} = \prod_{j=0}^{n-1} (Z_1 - \zeta_n^j Z_2) = Z_1^n - Z_2^n.$$

The characters of these forms are determined by their actions on lifts of s_n and $t_D = \begin{bmatrix} 0 & i \\ i & 0 \end{bmatrix}$, s'_n as above and $t'_D : (Z_1, Z_2) \mapsto (iZ_2, iZ_1)$. Easy calculations give $F_{1,D_n} \circ s'_n = F_{1,D_n}$ and $F_{1,D_n} \circ t'_D = -F_{1,D_n}$. Similar calculations for F_{2,D_n} and F_{3,D_n} give

$$\chi_{1,D_n} : s'_n \mapsto 1, \quad t'_D \mapsto -1,$$
$$\chi_{2,D_n} : s'_n \mapsto -1, \quad t'_D \mapsto i^n,$$
$$\chi_{3,D_n} : s'_n \mapsto -1, \quad t'_D \mapsto -i^n,$$
$$\chi_{1,D_n}^n = \chi_{2,D_n}^2 = \chi_{3,D_n}^2 = \chi_{D_n} : s'_n \mapsto 1, \quad t'_D \mapsto (-1)^n.$$

The characters χ_{2,D_n} and χ_{3,D_n} descend to D_n only when n is even. The other two characters descend for all n. The syzygy on the degenerate dihedral orbit-forms is

$$4F_{1,D_n}^n - F_{2,D_n}^2 + F_{3,D_n}^2 = 0.$$

For the Platonic groups, recall the table

	n_1	n_2	n_3	$\lvert G\rvert$	$\lvert\mathcal{O}_1\rvert$	$\lvert\mathcal{O}_2\rvert$	$\lvert\mathcal{O}_3\rvert$
Tetrahedron	3	3	2	12	4	4	6
Octahedron	4	3	2	24	6	8	12
Icosahedron	5	3	2	60	12	20	30

In each case, $|\mathcal{O}_2| = 2(|\mathcal{O}_1| - 2)$ and $|\mathcal{O}_3| = 3(|\mathcal{O}_1| - 2)$. Thus once F_1 is known, F_2 and F_3 are simply HF_1 and JF_1 respectively: they have the right degrees, and—excepting the tetrahedral case—invariant form-classes of those degrees are unique since orbits of those lengths are unique. In the tetrahedral case, where $|\mathcal{O}_2| = |\mathcal{O}_1|$, we need to check that $HF_1 \neq F_1$. All Platonic orbits have even order, so all Platonic characters descend to the rotation groups and will be defined there.

The tetrahedral vertices in S^2, $\{(\pm1, \pm1, \pm1)/\sqrt{3} : xyz > 0\}$, project stereographically to the orbit $\mathcal{O}_1 = \{\pm(1+i)/(\sqrt{3}-1), \pm(1-i)/(\sqrt{3}+1)\}$, whose form (3.2.1) is a product of four linear factors working out to

$$F_{1,T} = Z_1^4 - 2i\sqrt{3}Z_1^2 Z_2^2 + Z_2^4.$$

Consequently,

$$F_{2,T} = HF_{1,T} = Z_1^4 + 2i\sqrt{3}Z_1^2 Z_2^2 + Z_2^4$$

up to constant multiple. This is distinct from $F_{1,T}$ as needed, and in hindsight had to be its complex conjugate since the tetrahedral face-centers are the conjugates of the vertices. Next,

$$F_{3,T} = JF_{1,T} = Z_1 Z_2 (Z_1^4 - Z_2^4)$$

again up to constant multiple. From now on, constant factors will be discarded without comment in computing form classes. The respective characters are

$$\begin{aligned}
\chi_{1,T} &: s_T \mapsto 1, \quad t_T \mapsto \zeta_3, \\
\chi_{2,T} &= \chi_{1,T}^2 = \chi_{1,T}^{-1}, \\
\chi_{3,T} &= \chi_{1,T}^3 = 1, \\
\chi_{1,T}^3 = \chi_{2,T}^3 &= \chi_{3,T}^2 = \chi_T = 1.
\end{aligned}$$

The syzygy on the degenerate tetrahedral orbit-forms is

$$F_{1,T}^3 - F_{2,T}^3 + 12i\sqrt{3}F_{3,T}^2 = 0.$$

The octahedral vertices project to $\mathcal{O}_1 = \{0, \infty, \pm1, \pm i\}$ with corresponding form (3.2.1) working out to

$$F_{1,O} = Z_1 Z_2 (Z_1^4 - Z_2^4).$$

Of course, this is $F_{3,T}$ since the octahedral vertices are the tetrahedral midedges. For similar reasons (Exercise 3.4.1), $F_{1,O} = F_{1,D_4} F_{3,D_4}$. Using the machinery again,

$$F_{2,O} = HF_{1,O} = Z_1^8 + 14 Z_1^4 Z_2^4 + Z_2^8$$

and

$$F_{3,O} = JF_{1,O} = Z_1^{12} - 33Z_1^8Z_2^4 - 33Z_1^4Z_2^8 + Z_2^{12}.$$

The respective characters are

$$\chi_{1,O} : s_O \mapsto -1, \quad t_O \mapsto 1,$$
$$\chi_{2,O} = \chi_{1,O}^2 = 1,$$
$$\chi_{3,O} = \chi_{1,O}^3 = \chi_{1,O},$$
$$\chi_{1,O}^4 = \chi_{2,O}^3 = \chi_{3,O}^2 = \chi_O = 1.$$

The syzygy on the degenerate octahedral forms is

$$108F_{1,O}^4 - F_{2,O}^3 + F_{3,O}^2 = 0.$$

For the icosahedron, the projected vertices are $\mathcal{O}_1 = \{0, \zeta_5^j t_I 0, \infty, \zeta_5^j t_I \infty : 0 \le j \le 4\}$ (this was Exercise 2.7.5). Recall that $t_I = \begin{bmatrix} -1 & g \\ g & 1 \end{bmatrix}$ where the golden ratio g is the positive root of $x^2 + x - 1 = 0$; the negative root is $-g^{-1} = -1 - g$, denoted $\tilde{g}$. Thus $t_I 0 = g$ and $t_I \infty = \tilde{g}$, so $\mathcal{O}_1 = \{0, \zeta_5^j g, \infty, \zeta_5^j \tilde{g} : 0 \le j \le 4\}$ whose corresponding form (3.2.1) is

$$\begin{aligned} F_{1,I} &= Z_1 Z_2 \prod_{j=0}^{4} (Z_1 - \zeta_5^j g Z_2) \prod_{j=0}^{4} (Z_1 - \zeta_5^j \tilde{g} Z_2) \\ &= Z_1 Z_2 (Z_1^5 - g^5 Z_2^5)(Z_1^5 - \tilde{g}^5 Z_2^5) \\ &= Z_1 Z_2 (Z_1^{10} - (g^5 + \tilde{g}^5) Z_1^5 Z_2^5 + g^5 \tilde{g}^5 Z_2^{10}). \end{aligned}$$

Since $g, \tilde{g}$ are the roots of $x^2 = 1 - x$, repeated substitution of $1 - g$ for g^2 gives $g^5 = 5g - 3$ and similarly $\tilde{g}^5 = 5\tilde{g} - 3$; since $g + \tilde{g} = -1$, $g^5 + \tilde{g}^5 = 5(g + \tilde{g}) - 6 = -11$; and since $g\tilde{g} = -1$, $g^5\tilde{g}^5 = -1$. Thus the form for $\mathcal{O}_1$ simplifies to

$$F_{1,I} = Z_1 Z_2 (Z_1^{10} + 11 Z_1^5 Z_2^5 - Z_2^{10}).$$

Applying the covariants gives

$$F_{2,I} = HF_{1,I} = -(Z_1^{20} + Z_2^{20}) + 228(Z_1^{15}Z_2^5 - Z_1^5Z_2^{15}) - 494Z_1^{10}Z_2^{10}$$

and

$$F_{3,I} = JF_{1,I} = (Z_1^{30} + Z_2^{30}) + 522(Z_1^{25}Z_2^5 - Z_1^5Z_2^{25}) - 10005(Z_1^{20}Z_2^{10} + Z_1^{10}Z_2^{20}).$$

By Exercise 3.4.4, all three associated characters are trivial, as is the character $\chi_{1,I}^5 = \chi_{2,I}^3 = \chi_{3,I}^2 = \chi_I$. The syzygy on the degenerate icosahedral forms is

$$1728F_{1,I}^5 - F_{2,I}^3 - F_{3,I}^2 = 0.$$

Exercises

3.4.1. Explain geometrically why $F_{1,O} = F_{1,D_4} F_{3,D_4}$.

3.4.2. Verify the calculations of the degenerate orbit-forms $F_{i,\Gamma}$ in this section.

3.4.3. (a) Show that any finite subgroup of $\mathbf{C}^*$ is cyclic.
(b) Let G be a finite group and $\chi : G \longrightarrow \mathbf{C}^*$ a character. Show that χ descends to a character on G/K for some $K \triangleleft G$ such that G/K is cyclic.

3.4.4. This exercise shows that the only character of the group A_n for $n \geq 5$ is the trivial character, $\chi(\sigma) = 1$ for all $\sigma \in A_n$.
(a) First some general group theory: If g_1, g_2 are elements of a group G, their **commutator** is $[g_1, g_2] = g_1 g_2 g_1^{-1} g_2^{-1}$, which is 1 exactly when g_1 and g_2 commute. The **commutator subgroup** of G is $[G, G] = \langle [g_1, g_2] : g_1, g_2 \in G \rangle$, the subgroup of G generated by commutators. Show that $[G, G]$ is normal in G by exhibiting any element $\gamma [g_1, g_2] \gamma^{-1}$ as a product of commutators. Show that for any normal subgroup $K \triangleleft G$, G/K is abelian if and only if $[G, G] \subset K$.
(b) Working in the symmetric group S_n (with right-to-left multiplication), show that all even permutations are products of 3-cycles (it suffices to show this for $(2\,3)(1\,2)$ and $(3\,4)(1\,2)$). Show that all 3-cycles $(a\,b\,c)$ are conjugate in A_n to $(1\,2\,3)$ by evaluating the product

$$\begin{pmatrix} a & b & c & d & e \\ 1 & 2 & 3 & 4 & 5 \end{pmatrix} \begin{pmatrix} a & b & c \end{pmatrix} \begin{pmatrix} 1 & 2 & 3 & 4 & 5 \\ a & b & c & d & e \end{pmatrix}$$

and explaining why the conjugating permutation may be taken as even. Thus, to show that a normal subgroup $K \triangleleft A_n$ is all of A_n, it suffices to exhibit a single 3-cycle in K.
(c) Now suppose $\chi : A_n \longrightarrow \mathbf{C}^*$ is a character. Let $K = \ker(\chi) \triangleleft A_n$. Use the preceding exercise and part (a) to explain why K contains the commutator subgroup $[A_n, A_n]$ and in particular contains $(354)(132)(345)(123)$. Complete the proof that χ is trivial.

3.4.5. Verify the calculations of the degenerate orbit-form characters $\chi_{i,\Gamma}$ in this section.

3.4.6. Verify the syzygies asserted in the section.

3.4.7. For each rotation group Γ, for what values of e_1, e_2, e_3, e is the general invariant form (3.2.3) **totally** invariant, meaning its character on Γ' is trivial?

5. Invariant algebraic mappings

Let Γ be any of the finite rotation groups, viewed as acting on $\mathbf{P}^1(\mathbf{C})$. Recall that an algebraic mapping $f = [G : H] : \mathbf{P}^1(\mathbf{C}) \longrightarrow \mathbf{P}^1(\mathbf{C})$ is called Γ-invariant if $f \circ \gamma = f$ for all $\gamma \in \Gamma$, and this condition is equivalent to G and H being Γ'-invariant with the same character.

This section gives a simple description of Γ-invariant algebraic mappings. A technical lemma is required first to dispense with certain irksome possibilities. Call a nonconstant form F **exceptional** if $F = F_1^{e_1} F_2^{e_2} F_3^{e_3}$ with $0 \le e_i < n_i$ for $i = 1, 2, 3$ and $\deg(F)$ is divisible by $|\Gamma|$. (As before, take $F_3 = 1$ for the cyclic group C_n to make the notation uniform.) Thinking about zero-sets shows that every Γ'-invariant form of degree $|\Gamma|$ is a full orbit-form or an exceptional form. Similarly, every Γ'-invariant form of degree divisible by $|\Gamma|$ is a product of full orbit-forms possibly times an exceptional form. These are Exercise 3.5.1.

(3.5.1) LEMMA. *Every exceptional form has degree $|\Gamma|$. No exceptional form has character χ_Γ. No two exceptional forms share the same character.*

PROOF. Let $F = F_1^{e_1} F_2^{e_2} F_3^{e_3}$ be exceptional. Thus $\sum_i e_i |\mathcal{O}_i|$ is a positive multiple of $|\Gamma|$. Consider the table in Figure 3.5.1.

	$\lvert\mathcal{O}_1\rvert$	$\lvert\mathcal{O}_2\rvert$	$\lvert\mathcal{O}_3\rvert$	$\lvert\Gamma\rvert$	χ_1	χ_2	χ_3	χ_Γ
C_n	1	1	0	n	$s'_n \mapsto \zeta_{2n}$	χ_1^{-1}	1	$s'_n \mapsto -1$
D_n	2	n	n	$2n$	$s'_n \mapsto 1$ $t'_D \mapsto -1$	$s'_n \mapsto -1$ $t'_D \mapsto i^n$	$s'_n \mapsto -1$ $t'_D \mapsto -i^n$	$s'_n \mapsto 1$ $t'_D \mapsto (-1)^n$
Γ_T	4	4	6	12	$s_T \mapsto 1$ $t_T \mapsto \zeta_3$	χ_1^{-1}	1	1
Γ_O	6	8	12	24	$s_O \mapsto -1$ $t_O \mapsto 1$	1	χ_1	1
Γ_I	12	20	30	60	1	1	1	1

Figure 3.5.1. Orbit-lengths and characters of the rotation groups

Exercise 3.5.2 asks for verification of the following casewise argument For the cyclic group C_n, the conditions on F are $e_1, e_2 < n$, and so $e_1 + e_2 = \deg(F)$, being a positive multiple of n, satisfies $e_1 + e_2 = n$. Thus $\chi_F = \chi_1^{e_1 - e_2} : s'_n \mapsto \zeta_{2n}^{e_1 - e_2} = \zeta_{2n}^{2e_1 - n} = -\zeta_n^{e_1}$; these characters are distinct for $e_1 = 1, \ldots, n-1$ and distinct from χ_{C_n}. For the dihedral group, the possibilities for (e_1, e_2, e_3) such that $0 \le e_1 < n$, $0 \le e_2, e_3 < 2$ and $2e_1 + ne_2 + ne_3$ is a positive multiple of $2n$ are $(0,1,1)$, $(n/2,1,0)$, $(n/2,0,1)$, where the last two assume n is even. The corresponding characters χ_F are again distinct and not χ_{D_n}. For the tetrahedral group, the possibilities are $(e_1,e_2,e_3) \in \{(2,1,0),(1,2,0)\}$, with corresponding characters $\chi_{1,T}$ and $\chi_{1,T}^{-1}$ distinct and nontrivial. For the octahedral group, the only possibility is $(e_1,e_2,e_3) = (2,0,1)$, and again the character is nontrivial. For the icosahedral group there are no possible (e_1,e_2,e_3).

The first statement of the lemma falls out from this enumeration, or see Exercise 3.5.3 for a simple direct proof. □

Now the classification is easy. First the simplest algebraic mappings:

(3.5.2) THEOREM. *The Γ-invariant algebraic mappings of degree $|\Gamma|$ are*

$$\{f = [G : H] : G \textit{ and } H \textit{ are distinct full orbit-forms}\}.$$

This follows from the more general result,

(3.5.3) THEOREM. *The Γ-invariant algebraic mappings are*

$$\{f = [G : H] : \deg(G) = \deg(H);\ G,\ H \textit{ are disjoint products of full orbit-forms}\}.$$

PROOF. Clearly any such f is Γ-invariant. For the converse, let f be a nonconstant Γ-invariant algebraic mapping. Then excepting its values on the degenerate orbits, f takes every value in $\mathbf{P}^1(\mathbf{C})$ on a multiunion of full Γ-orbits, so its degree is divisible by $|\Gamma|$. Thus by (3.2.3), $f = [G : H]$ with G and H each a product of full orbit-forms possibly times an exceptional form. But neither exceptional form can be present if G and H are to share the same character. □

Exercises

3.5.1. Show that every Γ'-invariant form of degree $|\Gamma|$ is a full orbit-form or an exceptional form. Show that every Γ'-invariant form of degree divisible by $|\Gamma|$ is a product of full orbit-forms possibly times an exceptional form.

3.5.2. Confirm the assertions about possible (e_1,e_2,e_3) and the corresponding characters for the various groups in the proof of Lemma 3.5.1.

3.5.3. Here is a direct argument that an exceptional form F must have degree $|\Gamma|$. It suffices to show that $\deg(F) < 2|\Gamma|$. Show that for an exceptional form F, $\deg(F) \le \sum_i (n_i - 1)|\mathcal{O}_i| = |\Gamma| \sum_i (1 - 1/n_i)$, where the sum is over degenerate orbits. Use (2.6.3) to finish the proof.

6. Invariant rational functions

Returning finally to the Riemann sphere $\widehat{\mathbf{C}}$, recall that a rational function $f = g/h : \widehat{\mathbf{C}} \longrightarrow \widehat{\mathbf{C}}$ is called Γ-invariant if $f \circ \gamma = f$ for all $\gamma \in \Gamma$. Under the correspondence between algebraic mappings and rational functions, Theorem 3.5.2 becomes

(3.6.1) THEOREM. *The Γ-invariant rational functions of degree $|\Gamma|$ are*

$$\left\{ f = \begin{bmatrix} a & b \\ c & d \end{bmatrix} \frac{(F_1^{n_1})_*}{(F_2^{n_2})_*} : \begin{bmatrix} a & b \\ c & d \end{bmatrix} \in \mathrm{PSL}_2(\mathbf{C}) \right\}.$$

Any two such functions f_1 and f_2 are fractional linear transformations of one another.

PROOF. The rational function f corresponding to $[G : H]$, where G and H are full orbit-forms $G = aF_1^{n_1} + bF_2^{n_2}$ and $H = cF_1^{n_1} + dF_2^{n_2}$, is

$$f = \frac{G_*}{H_*} = \frac{a(F_1^{n_1})_* + b(F_2^{n_2})_*}{c(F_1^{n_1})_* + d(F_2^{n_2})_*} = \begin{bmatrix} a & b \\ c & d \end{bmatrix} \frac{(F_1^{n_1})_*}{(F_2^{n_2})_*}.$$

Consequently, any two such functions f_1 and f_2, both being fractional linear transformations of $(F_1^{n_1})_*/(F_2^{n_2})_*$, are fractional linear transformations of one another. □

A corollary is that all Γ-invariant rational functions f of degree $|\Gamma|$ generate the same field of rational functions $\mathbf{C}(f)$. Thus Theorem 3.5.3 gives

(3.6.2) THEOREM. *The field $\mathbf{C}(Z)^\Gamma$ of Γ-invariant rational functions is $\mathbf{C}(f)$ for any Γ-invariant f of degree $|\Gamma|$.*

PROOF. The field is

$$\begin{aligned} \mathbf{C}(Z)^\Gamma &= \{G_*/H_* : \deg(G) = \deg(H);\ G,\ H \text{ are disjoint products} \\ &\qquad \text{of full orbit-forms}\} \\ &= \left\{ \prod_i (G_i)_*/(H_i)_* : G_i \text{ and } H_i \text{ are full orbit-forms} \right\} \\ &= \left\{ \prod_i \begin{bmatrix} a_i & b_i \\ c_i & d_i \end{bmatrix} \frac{(F_1^{n_1})_*}{(F_2^{n_2})_*} : \begin{bmatrix} a_i & b_i \\ c_i & d_i \end{bmatrix} \in \mathrm{PSL}_2(\mathbf{C}) \right\}. \end{aligned}$$

Each element of this set is rational in $(F_1^{n_1})_*/(F_2^{n_2})_*$, which is rational in f. Thus $\mathbf{C}(Z)^\Gamma \subset \mathbf{C}(f)$. The other containment is obvious. □

For the cyclic group C_n, the obvious generator for $\mathbf{C}(Z)^{C_n}$ is $f_{C_n} = Z^n$. For the other groups, a normalized generator f_Γ is specified by the noncanonical conditions that $f_\Gamma(\mathcal{O}_1) = \infty$, $f_\Gamma(\mathcal{O}_2) = 0$, $f_\Gamma(\mathcal{O}_3) = 1$, taken from Klein. (See Exercise 3.6.2.) Thus for some scalars a and b,

$$f_\Gamma = a\frac{(F_2^{n_2})_*}{(F_1^{n_1})_*} = 1 - b\frac{(F_3^{n_3})_*}{(F_1^{n_1})_*}.$$

Equivalently, $F_1^{n_1} - aF_2^{n_2} - bF_3^{n_3} = 0$, so that for each group Γ the scalars a and b can be read off from the syzygy among the degenerate orbit-forms, though we only need a. The results are

$$f_{D_n} = \frac{(F_{2,D_n}^2)_*}{4(F_{1,D_n}^n)_*} = \frac{(Z^n+1)^2}{4Z^n},$$

$$f_T = \frac{(F_{2,T}^3)_*}{(F_{1,T}^3)_*} = \left(\frac{Z^4 + 2i\sqrt{3}Z^2 + 1}{Z^4 - 2i\sqrt{3}Z^2 + 1}\right)^3,$$

$$f_O = \frac{(F_{2,O}^3)_*}{108(F_{1,O}^4)_*} = \frac{(Z^8 + 14Z^4 + 1)^3}{108Z^4(Z^4-1)^4},$$

$$f_I = \frac{(F_{2,I}^3)_*}{1728(F_{1,I}^5)_*} = \frac{(-Z^{20} + 228Z^{15} - 494Z^{10} - 228Z^5 - 1)^3}{1728Z^5(Z^{10} + 11Z^5 - 1)^5}.$$

Exercises

3.6.1. Let Γ be a rotation group acting on $\widehat{\mathbf{C}}$. The set of Γ-orbits is denoted $\Gamma\backslash\widehat{\mathbf{C}}$. The natural surjection

$$\mathrm{orb} : \widehat{\mathbf{C}} \longrightarrow \Gamma\backslash\widehat{\mathbf{C}} \qquad \text{where } \mathrm{orb}(p) = \mathcal{O}_p$$

induces a quotient topology on $\Gamma\backslash\widehat{\mathbf{C}}$, now called the **orbit space**, by the usual rule

$$S \subset \Gamma\backslash\widehat{\mathbf{C}} \text{ is open} \qquad \Longleftrightarrow \qquad \mathrm{orb}^{-1}S \subset \widehat{\mathbf{C}} \text{ is open.}$$

The map $f_\Gamma : \widehat{\mathbf{C}} \longrightarrow \widehat{\mathbf{C}}$ is constant on each orbit, so it induces a map $\hat{f}_\Gamma : \Gamma\backslash\widehat{\mathbf{C}} \longrightarrow \widehat{\mathbf{C}}$ making the following diagram commute.

$$\begin{array}{ccc} \widehat{\mathbf{C}} & & \\ {\scriptstyle \mathrm{orb}}\downarrow & \searrow^{f_\Gamma} & \\ \Gamma\backslash\widehat{\mathbf{C}} & \xrightarrow{\hat{f}_\Gamma} & \widehat{\mathbf{C}} \end{array}$$

Show that $\hat{f}_\Gamma$ is a homeomorphism as follows.
(a) Show that $\hat{f}_\Gamma$ is surjective.

(b) Show that $\hat{f}_\Gamma$ is injective. (Hint: if $\hat{f}_\Gamma(\mathcal{O}_p) = \hat{f}_\Gamma(\mathcal{O}_q)$ then no function in $\mathbf{C}(f_\Gamma)$ can take different values on $\mathcal{O}_p$ and $\mathcal{O}_q$. Find a function in $\mathbf{C}(f_\Gamma)$ that does so.)

(c) Show that $\hat{f}_\Gamma$ is continuous. (Recall from Chapter 1 that the meromorphic function f_Γ is continuous.)

(d) Show that $\hat{f}_\Gamma$ is open, meaning it takes open sets to open sets. (Hint: it suffices to show this for f_Γ, which is nonconstant analytic in local coordinates.)

3.6.2. Suppose f and g are rational functions related by a fractional linear transformation, i.e., $f = \begin{bmatrix} a & b \\ c & d \end{bmatrix} g$ for some $\begin{bmatrix} a & b \\ c & d \end{bmatrix} \in \mathrm{PSL}_2(\mathbf{C})$. Suppose further that f and g take distinct points p_1, p_2, p_3 to distinct points q_1, q_2, q_3. Show that $f = g$. Since for each rotation group Γ, all Γ-invariant rational functions f of degree $|\Gamma|$ are related by fractional linear transformations, this shows that any particular such f is determined by its values on three distinct orbits.

3.6.3. With the help of a symbolic algebra computer program, verify directly for some of the rotation groups Γ that f_Γ is invariant under the generators s and t of Γ.

7. Summary

For each rotation group Γ, the field $\mathbf{C}(Z)^\Gamma$ of Γ-invariant rational functions is generated by any member of degree $|\Gamma|$. A noncanonical generator is f_Γ, normalized (in the noncyclic case) to take the values ∞, 0, 1 at the vertices, face-centers, and mid-edges of the geometric figure corresponding to Γ. All other generators are fractional linear transformations of this one. Every Γ-invariant rational function, and in particular f_Γ, is the quotient of (dehomogenized) same-degree Γ'-invariant forms, which are easy to compute with general machinery.

CHAPTER 4

Inverses of the invariant functions

For each rotation group Γ, a normalized rational expression f_Γ of degree $|\Gamma|$ now generates the field of Γ-invariant functions in $\mathbf{C}(Z)$, i.e., $\mathbf{C}(f_\Gamma) = \mathbf{C}(Z)^\Gamma$. As a function, f_Γ maps the set of Γ-orbits in $\widehat{\mathbf{C}}$ to another copy of $\widehat{\mathbf{C}}$. This chapter discusses the problem of recovering an orbit from its f_Γ-value. Algebraically, the problem is to invert f_Γ: letting $W = f_\Gamma(Z)$, reconstruct Z as a multiple-valued expression in W.

In the cyclic case this is easy enough. Since $W = f_{C_n}(Z) = Z^n$, the inverse is the radical $Z = \sqrt[n]{W}$. The dihedral case isn't much harder since the quadratic formula uses a square root to solve the relation $W = f_{D_n}(Z) = (Z^n + 1)^2/(4Z^n)$ for Z^n in terms of W, after which an nth root suffices again. But the Platonic invariants become progressively more complicated, until inverting them by radicals is infeasible ad hoc and perhaps not even possible.

The environment for discussing this sort of problem is Galois theory, a subject whose sheer elegance makes it a crown jewel of undergraduate mathematics. This chapter begins with general material on fields and Galois theory, developing ideas that explain when and how to invert by radicals. Certain useful polynomials called resolvents, presented later in the chapter, further systematize and illuminate the calculations. For all rotation groups Γ but the icosahedral group, inverting f_Γ reduces to finding the zeros of a succession of polynomials by field operations and nth roots. For the icosahedral group, inverting f_I can not be done by radicals, but instead is equivalent to solving a specific quintic polynomial with coefficients in the field $\mathbf{C}(W)$. This special quintic, called the Brioschi resolvent, will play a prominent role in the rest of the book.

Recommended reading: Among the infinitely many good books on Galois theory, Stewart [Ste] is a successful undergraduate text and Jacobson [Ja I] presents the material nicely. Lang's [La] chapters on field theory and Galois theory are quite accessible. The specific calculations in this chapter are from

Chapter I.4 of Klein [Kl] and Chapter XIII of Dickson [Di].

1. Fields and polynomials

Let $\mathbf{k}$ be a field. The natural ring homomorphism $\varphi : \mathbf{Z} \longrightarrow \mathbf{k}$ such that $\varphi(1_\mathbf{Z}) = 1_\mathbf{k}$ (see Exercise 4.1.1) has an integral domain as its image. By the First Isomorphism Theorem for rings (Exercise 4.1.2), $\ker(\varphi)$ is an ideal I such that $\mathbf{Z}/I$ is an integral domain, so I is a prime ideal in $\mathbf{Z}$ (see Exercise 4.1.3). Such ideals are $\{0\}$ and $p\mathbf{Z}$ for p prime. If $I = \{0\}$ then $\mathbf{k}$ contains a copy of the ring $\mathbf{Z}$ and therefore contains a copy of the field $\mathbf{Q}$; otherwise $\mathbf{k}$ contains a copy of the field $\mathbf{F}_p = \mathbf{Z}/p\mathbf{Z}$ for some prime p. The copy of $\mathbf{Q}$ or $\mathbf{F}_p$ in $\mathbf{k}$ is contained in all subfields of $\mathbf{k}$ (indeed, it is their intersection) and is called the **prime subfield** of $\mathbf{k}$. When the prime subfield is (isomorphically) $\mathbf{Q}$, $\mathbf{k}$ has **characteristic zero**, and for any nonzero $x \in \mathbf{k}$ and $n \in \mathbf{Z}^+$, $x + x + \cdots + x$ (n summands) is nonzero. When the prime subfield is $\mathbf{F}_p$, $\mathbf{k}$ has **characteristic** p, and for any $x \in \mathbf{k}$, $x + x + \cdots + x$ (p summands) equals 0.

For any field $\mathbf{k}$, the ring $\mathbf{k}[T]$ of polynomials over $\mathbf{k}$ in the unknown T contains $\mathbf{k}$ as a subring, and the units (invertible elements) of $\mathbf{k}[T]$ are $\mathbf{k}^*$. The polynomial ring $\mathbf{k}[T]$ comes with the degree map $\deg : \mathbf{k}[T] \longrightarrow \mathbf{N} \cup \{-\infty\}$ where $\deg(0) = -\infty$ and $\deg(\sum_{i=0}^{d} a_i T^i) = d$ assuming $a_d \neq 0$. In fact $\mathbf{k}[T]$ is a Euclidean ring under this map, meaning that for all $g, h \in \mathbf{k}[T]$ with h nonzero, the relation $\deg(g) \leq \deg(gh)$ holds and there exist $q, r \in \mathbf{k}[T]$ satisfying the division formula $g = qh + r$, $\deg(r) < \deg(h)$. By general ring theory it follows that $\mathbf{k}[T]$ is a unique factorization domain (Exercise 4.1.6 or Exercise 4.1.7). The Euclidean algorithm applies in $\mathbf{k}[T]$ (Exercise 4.1.8): given polynomials g and h, there exist G and H such that

$$Gg + Hh = \gcd(g, h).$$

Exercises

4.1.1. Let $\mathbf{k}$ be a field. Prove that the condition $\varphi(1_\mathbf{Z}) = 1_\mathbf{k}$ defines a ring homomorphism $\varphi : \mathbf{Z} \longrightarrow \mathbf{k}$. Use the homomorphism φ to define integer multiplication on $\mathbf{k}$, $\cdot : \mathbf{Z} \times \mathbf{k} \longrightarrow \mathbf{k}$, by $n \cdot x = \varphi(n)x$. (This makes $\mathbf{k}$ a **Z-module**.) Show that for all $n \in \mathbf{N}$ and $x \in \mathbf{k}$, $n \cdot x = x + \cdots + x$ (n summands). Thus $\mathbf{k}$ has characteristic zero if 0 is the only integer n such that $n \cdot x = 0$ for all $x \in \mathbf{k}$, and $\mathbf{k}$ has characteristic p if $p \cdot x = 0$ for all $x \in \mathbf{k}$.

4.1.2. State and prove the First Isomorphism Theorem for rings.

4.1.3. Prove that the quotient R/I of a commutative ring with unit by an ideal gives an integral domain if and only if I is prime (if a product ab lies in I then so does a or b), and gives a field if and only if I is maximal (the only ideal properly containing I is all of R).

4.1.4. Let p be a prime. Show that $p \mid \binom{p}{j}$ for $0 < j < p$, so that in any field $\mathbf{k}$ of characteristic p, $(x+y)^p = x^p + y^p$ for all $x, y \in \mathbf{k}$. Show by induction that for any positive integer n, $(x+y)^{p^n} = x^{p^n} + y^{p^n}$ for all $x, y \in \mathbf{k}$.

4.1.5. Show that the quotient q and remainder r in the division formula for $\mathbf{k}[T]$ are unique.

4.1.6. (a) Show that every Euclidean ring is a principal ideal domain.
(b) Show that every principal ideal domain is a unique factorization domain. (This is fairly substantial and sometimes omitted in a first algebra course because the next exercise is easier. It is worth looking up if you haven't seen it before and have time.)

4.1.7. Show directly that every Euclidean ring is a unique factorization domain. (See, for example, Herstein [He].)

4.1.8. Show that the Euclidean algorithm applies in $\mathbf{k}[T]$.

2. Algebraic extensions

A **field extension** is a containment $\mathbf{k} \subset \mathbf{K}$ of two fields, written $\mathbf{K}/\mathbf{k}$. Since $\mathbf{K}$ is a vector space over $\mathbf{k}$, defining the **degree** $[\mathbf{K} : \mathbf{k}]$ of the extension as $\dim_{\mathbf{k}}(\mathbf{K})$ is meaningful. The extension is called **finite** when its degree is finite. (See Exercise 4.2.1.)

Given a field $\mathbf{k}$ and an element r of some field $\mathbf{K}$ containing $\mathbf{k}$, $\mathbf{k}(r)/\mathbf{k}$ is a field extension, where $\mathbf{k}(r)$ is the intersection of all subfields of $\mathbf{K}$ containing $\mathbf{k}$ and r, or equivalently $\mathbf{k}(r)$ is the rational expressions generated by $\mathbf{k}$ and r. Exercises 4.2.2, 4.2.3, and 4.2.5 establish the following facts: The extension $\mathbf{k}(r)/\mathbf{k}$ is finite if and only if r is a root of some nonzero polynomial $h \in \mathbf{k}[T]$, in which case r is called **algebraic** over $\mathbf{k}$. Any $r \in \mathbf{K}$ algebraic over $\mathbf{k}$ is in fact a root of a unique monic polynomial $m \in \mathbf{k}[T]$ of least degree, called the **minimal polynomial of** r. The minimal polynomial of r is irreducible, and $[\mathbf{k}(r) : \mathbf{k}] = \deg(r)$. More generally, a field extension $\mathbf{K}/\mathbf{k}$ is called **algebraic** if every element $r \in \mathbf{K}$ is algebraic over $\mathbf{k}$. Every finite extension is algebraic, but the converse is not true.

Let $\mathbf{k}$ be a field and $h \in \mathbf{k}[T]$ a polynomial. The field $\mathbf{K}$ is a **splitting field** over $\mathbf{k}$ of h, written

$$\mathbf{K} = \mathrm{spl}_{\mathbf{k}}(h),$$

if $\mathbf{K} = \mathbf{k}(r_1, \ldots, r_k)$ (meaning $\mathbf{K}$ is the field generated by $\mathbf{k}$ and $r_1, \ldots, r_k$), where $h = a\prod_{i=1}^{k}(T - r_i)$ in $\mathbf{K}[T]$. In other words, $\mathbf{K}$ is generated over $\mathbf{k}$ by a complete set of roots of h, or equivalently, h factors down to linear terms in $\mathbf{K}$ but not in any proper subfield of $\mathbf{K}$ containing $\mathbf{k}$. For example, $\mathbf{Q}(\sqrt[3]{2}, \zeta_3\sqrt[3]{2}, \zeta_3^2\sqrt[3]{2}) = \mathbf{Q}(\sqrt[3]{2}, \zeta_3) \subset \mathbf{C}$ is a splitting field over $\mathbf{Q}$ of the polynomial $h = T^3 - 2$, where $\sqrt[3]{2}$ is the real cube root of 2 and as usual $\zeta_3 = e^{2\pi i/3}$.

In the preceding example, working in the large field $\mathbf{C}$ containing all roots of the polynomial h made exhibiting $\mathrm{spl}_{\mathbf{Q}}(h)$ an easy matter. But since $\mathbf{C}$ is extrinsic to the original objects $\mathbf{Q}$ and h, this doesn't feel entirely satisfactory. Using intrinsic methods, one can start from a field $\mathbf{k}$ and a nonconstant polynomial $h \in \mathbf{k}[T]$, and construct a splitting field. First construct a **root field** extension $\mathbf{k}(r)/\mathbf{k}$, where r is a root of h, as follows. Let m be a nonconstant irreducible factor of h. Consider the ring

$$\tilde{\mathbf{K}} = \mathbf{k}[T]/m\mathbf{k}[T].$$

This is a field because $m\mathbf{k}[T]$ is a maximal ideal in $\mathbf{k}[T]$ (Exercise 4.2.6). In fact $\tilde{\mathbf{K}}$ is a template for the desired $\mathbf{k}(r)$ since working in $\tilde{\mathbf{K}}$ while suppressing coset notation gives $m(T) = 0$, showing T is a root of m and therefore of h.

The formal construction of $\mathbf{k}(r)$ is a bit finicky. The map $\sigma : \mathbf{k} \longrightarrow \tilde{\mathbf{K}}$ such that $x^\sigma = x + m\mathbf{k}[T]$ (in field theory the action of a map is often indicated by a superscript in this fashion) is an injective homomorphism, also called an **embedding**; so $\tilde{\mathbf{K}}$ extends an isomorphic copy of $\mathbf{k}$. Constructing a corresponding extension of $\mathbf{k}$ itself—and thus replacing the embedding σ by a true inclusion—amounts to set-theoretic bookkeeping. Let $\mathbf{K}$ be a set of symbols including the elements of $\mathbf{k}$ such that the bijection $\sigma : \mathbf{k} \longrightarrow \mathbf{k} + m\mathbf{k}[T]$ extends to a bijection $\sigma : \mathbf{K} \longrightarrow \tilde{\mathbf{K}}$. Let r be the symbol in $\mathbf{K}$ such that $r^\sigma = T + m\mathbf{k}[T]$. Define operations on $\mathbf{K}$ to make the bijection σ a homomorphism; these extend the operations on $\mathbf{k}$ and make $\mathbf{K}$ a field, so $\mathbf{K}/\mathbf{k}$ is a field extension. Since $\tilde{\mathbf{K}}$ is generated by $T + m\mathbf{k}[T]$ over $\mathbf{k}^\sigma$, $\mathbf{K}$ is generated by r over $\mathbf{k}$, i.e., $\mathbf{K} = \mathbf{k}(r)$. Finally, if $m = \sum_{i=0}^{d} a_i T^i$, compute that

$$\begin{aligned} m(r)^\sigma &= (\sum_{i=0}^{d} a_i r^i)^\sigma = \sum_{i=0}^{d} a_i^\sigma (r^\sigma)^i = \sum_{i=0}^{d} (a_i + m\mathbf{k}[T])(T + m\mathbf{k}[T])^i \\ &= \sum_{i=0}^{d} (a_i T^i + m\mathbf{k}[T]) = m + m\mathbf{k}[T] = m\mathbf{k}[T] = 0_{\tilde{\mathbf{K}}} \end{aligned}$$

to conclude that $m(r) = 0_{\mathbf{K}}$, so $h(r) = 0_{\mathbf{K}}$.

This construction is unique up to isomorphism when h is irreducible. It shows that the field $\mathbf{k}(r)$ is simply the polynomial ring $\mathbf{k}[r]$ manipulated subject to the rule $m(r) = 0$ to eliminate all polynomials of degree $\deg(m)$ or higher. The most familiar example of this is given by the field $\mathrm{spl}_{\mathbf{R}}(T^2 + 1)$, otherwise known as $\mathbf{C}$. If i denotes a root of $T^2 + 1$ then the other root is $-i$, so the splitting field is $\mathbf{R}(i)$; manipulations with the rule $i^2 = -1$ recover the fact that $\mathbf{C} = \{x + iy : x, y \in \mathbf{R}\}$.

Iterating the construction yields $\mathrm{spl}_{\mathbf{k}}(h)$. Certainly $T - r$ divides h in $\mathbf{k}(r)[T]$. Set $h_2 = h/(T - r)$. If h_2 factors down to linear terms in $\mathbf{k}(r)[T]$ then h has a full set of roots in $\mathbf{k}(r)$. Otherwise, replace $\mathbf{k}$ and h by $\mathbf{k}(r)$ and h_2 respectively and repeat the root adjunction process to get an extension $\mathbf{k}(r, r_2)/\mathbf{k}$ containing at least two roots of h. Continuing this procedure eventually gives the splitting field. Many algebra books (see, for example, [La]) prove that "the" splitting field is indeed unique up to isomorphism, so we will not do so here. Building on the root adjunction process just described also constructs the **algebraic closure** of $\mathbf{k}$, the smallest field $\overline{\mathbf{k}}$ containing $\mathbf{k}$ such that the irreducible polynomials in $\overline{\mathbf{k}}[T]$ have degree 1, i.e., every polynomial in $\overline{\mathbf{k}}[T]$ factors down to linear terms. The algebraic closure is also unique up to isomorphism. Again, [La] gives the details.

Returning to the situation of a field $\mathbf{K}$ containing $\mathbf{k}$, an element $r \in \mathbf{k}$ is called **transcendental** over $\mathbf{k}$ if it is not algebraic over $\mathbf{k}$, that is, it is not a root of any polynomial $h \in \mathbf{k}[T]$. It follows that r satisfies no rational relation over $\mathbf{k}$ and the field $\mathbf{k}(r)$ consists of all formal rational expressions in r. Our familiar example of this is the field $\mathbf{C}(Z)$ of rational functions on $\widehat{\mathbf{C}}$.

Exercises

4.2.1. If $\mathbf{K}/\mathbf{k}$ is a field extension, show that $\mathbf{K}$ is a vector space over $\mathbf{k}$. If also $\mathbf{L}/\mathbf{K}$ is a field extension, show that $[\mathbf{L} : \mathbf{k}] = [\mathbf{L} : \mathbf{K}][\mathbf{K} : \mathbf{k}]$, meaning that if one side is infinite then so is the other, and if both sides are finite then they agree.

4.2.2. Let $\mathbf{k}$ be a field and r be an element of some field $\mathbf{K}$ containing $\mathbf{k}$. Show that if the extension $\mathbf{k}(r)/\mathbf{k}$ is finite then r is a root of some nonzero polynomial $h \in \mathbf{k}[T]$. (Hint: consider the set $\{1, r, r^2, \dots\}$.) For the converse, first show that the polynomials in $\mathbf{k}[T]$ to which r is a root form an ideal I. Consequently, if r is a root of some nonzero polynomial $h \in \mathbf{k}[T]$ then r is a root of a unique monic polynomial $m \in \mathbf{k}[T]$ of least degree. Show that this minimal polynomial of r

is irreducible, and therefore I is maximal. Show that consequently $\mathbf{k}(r) = \mathbf{k}[r]$ (consider the natural surjection $\mathbf{k}[T] \longrightarrow \mathbf{k}[r]$), and the extension $\mathbf{k}(r)/\mathbf{k}$ is therefore finite.

4.2.3. Show that every finite extension is algebraic.

4.2.4. This exercise describes all finite fields. Justify the statements as necessary. Let p be a prime and n a positive integer. Set $q = p^n$ and let $\mathbf{F}_q = \mathrm{spl}_{\mathbf{F}_p}(h)$ where $h = T^q - T \in \mathbf{F}_p[T]$. Show that the roots of h in $\mathbf{F}_q$ form a subfield, by verifying that 0 and 1 are roots, and if x and y are roots then so are $x+y$, $-x$, xy and x^{-1}, assuming $x \neq 0$ for the last case. (Exercise 4.1.4 will help. Checking that $-x$ is a root requires separate inspection of the cases $p = 2$ and p odd.) This subfield must be all of $\mathbf{F}_q$ by definition of splitting field. The roots of h are distinct since h and its derivative $h' = -1$ share no roots (see for example [Ja I]—polynomial derivatives and their properties can be developed purely algebraically with no reference to a limit process particular to the fields $\mathbf{R}$ and $\mathbf{C}$). Thus $\mathbf{F}_q$ has q elements, exhibiting a field of any prime power order.

Conversely, let $\mathbf{F}$ be any finite field. It must have prime subfield $\mathbf{F}_p$ for some p, and being a vector space over $\mathbf{F}_p$, it must contain $p^n = q$ elements for some positive n. Every $x \in \mathbf{F}^*$ is a root of the polynomial $T^{q-1} - 1$, so every $x \in \mathbf{F}$ is a root of $h = T^q - T$ and $\mathbf{F} = \mathrm{spl}_{\mathbf{F}_p}(h) = \mathbf{F}_q$ up to isomorphism. Thus the finite fields just constructed are all possible finite fields.

Working in a fixed algebraic closure of $\mathbf{F}_p$, show that $\mathbf{F}_q \subset \mathbf{F}_{q'}$ if and only if q' is a power of q. So, for example, $\mathbf{F}_9$ sits in $\mathbf{F}_{81}$ but not in $\mathbf{F}_{27}$.

4.2.5. Let p be a prime and consider the tower of fields

$$\mathbf{F}_p \subset \mathbf{F}_{p^2} \subset \mathbf{F}_{p^4} \subset \mathbf{F}_{p^8} \subset \cdots$$

Let $\mathbf{F} = \cup_i \mathbf{F}_{p^{2^i}}$ be the union of fields in the tower. Show that $\mathbf{F}/\mathbf{F}_p$ is an algebraic extension that is not finite.

4.2.6. If $m \in \mathbf{k}[T]$ is irreducible, show that $m\mathbf{k}[T]$ is a maximal ideal in $\mathbf{k}[T]$. (Apply the Euclidean algorithm to m and g for any $g \notin m\mathbf{k}[T]$.)

4.2.7. Let $\mathbf{k}$ be a field and $m \in \mathbf{k}[T]$ be an irreducible polynomial. Show that $\deg(m) \leq [\mathrm{spl}_{\mathbf{k}}(m) : \mathbf{k}]$ and $[\mathrm{spl}_{\mathbf{k}}(m) : \mathbf{k}] \mid \deg(m)!$.

4.2.8. **Gauss' Lemma** (to be proved in Exercise 4.4.2) says that if the polynomial $h \in \mathbf{Z}[T]$ factors in $\mathbf{Q}[T]$, it in fact factors in $\mathbf{Z}[T]$. **Eisenstein's criterion** says that if $m = \sum_{i=0}^{d} a_i T^i \in \mathbf{Z}[T]$ and there exists a prime p such that $p \mid a_0, \ldots, p \mid a_{d-1}, p \nmid a_d, p^2 \nmid a_0$, then m is irreducible in $\mathbf{Z}[T]$ and therefore irreducible in $\mathbf{Q}[T]$. Let p be a prime and consider the **cyclotomic** polynomial $m = 1+T+T^2+\cdots+T^{p-1}$. Use the geometric sum formula to show that $m(T+1)$ is irreducible in $\mathbf{Q}[T]$, and therefore so is m. Show that m is the minimal polynomial over $\mathbf{Q}$ of ζ_p. Describe the splitting field $\mathrm{spl}_{\mathbf{Q}}(m)$. What is $[\mathrm{spl}_{\mathbf{Q}}(m) : \mathbf{Q}]$? Express $1/\zeta_p$ as a polynomial in ζ_p with rational coefficients.

4.2.9. Let $r = \sqrt[3]{2+\sqrt{5}} + \sqrt[3]{2-\sqrt{5}}$, taking real cube roots. Find the minimal polynomial $m \in \mathbf{Q}[T]$ of r over $\mathbf{Q}$ and the degree $[\mathbf{Q}(r) : \mathbf{Q}]$. (You may be surprised by the answer.)

3. Galois extensions

If $\mathbf{K}$ and $\mathbf{L}$ are fields containing the field $\mathbf{k}$ then a homomorphism $\sigma : \mathbf{K} \longrightarrow \mathbf{L}$ is called a **k-map** (or, is a **map over k**) if $x^\sigma = x$ for all $x \in \mathbf{k}$, i.e., σ restricts to the identity on $\mathbf{k}$. If $\sigma : \mathbf{K} \longrightarrow \mathbf{L}$ is a $\mathbf{k}$-map, and $h \in \mathbf{k}[T]$ is a polynomial, and $r \in \mathbf{K}$ is a root of h, it follows that $h(r^\sigma) = h(r)^\sigma = 0_{\mathbf{K}}^\sigma = 0_{\mathbf{L}}$, so $r^\sigma \in \mathbf{L}$ is also a root of h; that is, $\mathbf{k}$-maps preserve roots of polynomials over $\mathbf{k}$. The automorphisms of $\mathbf{K}$ (isomorphisms from $\mathbf{K}$ to $\mathbf{K}$) over $\mathbf{k}$ form a group, denoted $\mathrm{Aut}_{\mathbf{k}}(\mathbf{K})$. Composition of automorphisms is carried out left-to-right to make the formula $x^{\sigma\tau} = (x^\sigma)^\tau$ valid.

The algebraic extension $\mathbf{K}/\mathbf{k}$ is called **normal** if for every $\mathbf{k}$-map $\sigma : \mathbf{K} \longrightarrow \overline{\mathbf{k}}$ (which must be an injection—see Exercise 4.3.1), in fact $\mathbf{K}^\sigma = \mathbf{K}$. Such σ is therefore an isomorphism and lies in $\mathrm{Aut}_{\mathbf{k}}(\mathbf{K})$. Thus normality is a certain closure condition, analogous to how closure under conjugation in group theory defines a normal subgoup. The preceding paragraph shows that splitting field extensions are normal (Exercise 4.3.2). More generally (see [La]), an algebraic extension $\mathbf{K}/\mathbf{k}$ is normal exactly when any irreducible polynomial $m \in \mathbf{k}[T]$ with a root in $\mathbf{K}$ has all of its roots in $\mathbf{K}$. A simple example of a non-normal extension is $\mathbf{Q}(\sqrt[3]{2})/\mathbf{Q}$, where $\sqrt[3]{2}$ is the real cube root of 2. (Exercise 4.3.3 asks why this extension is not normal.)

A polynomial $h \in \mathbf{k}[T]$ is **separable** if it factors as a product of distinct linear factors in $\mathrm{spl}_{\mathbf{k}}(h)$, i.e., it has no multiple roots. If $\mathbf{K}/\mathbf{k}$ is an extension, the element $r \in \mathbf{K}$ is called **separable** over $\mathbf{k}$ if r is algebraic over $\mathbf{k}$ and its minimal polynomial $m \in \mathbf{k}[T]$ is separable. The algebraic extension $\mathbf{K}/\mathbf{k}$ is **separable** if each $r \in \mathbf{K}$ is separable over $\mathbf{k}$.

All algebraic extensions of fields of characteristic zero are separable. To see this, it suffices to show that for any field $\mathbf{k}$ of characteristic zero, each irreducible polynomial $m \in \mathbf{k}[T]$ has distinct roots. Indeed, if m has a repeated root r in $\mathrm{spl}_{\mathbf{k}}(m)$, then r is also a root of the derivative m'. (See Exercise 4.2.4 for a comment on this.) The ideal of polynomials of which r is a root is generated by m, so m', having lower degree than m, must be the zero polynomial. In characteristic zero, this forces m to be constant, contradiction.

All algebraic extensions of finite fields (Exercise 4.2.4) are also separable. Exercise 4.3.4 proves this. Therefore an inseparable extension must involve infinite fields of characteristic p. See Exercise 4.3.5 for an example.

Any splitting field extension of a separable polynomial is a separable extension. This is proved in many algebra books, e.g., [La].

A finite algebraic extension $\mathbf{K}/\mathbf{k}$ is called a **Galois** extension if it is normal and separable. In characteristic zero, any finite normal extension, and in particular any splitting field extension, is Galois. By the fact cited in the preceding paragraph, a splitting field extension of a separable polynomial is Galois in arbitrary characteristic. When $\mathbf{K}/\mathbf{k}$ is Galois, the automorphism group $\mathrm{Aut}_{\mathbf{k}}(\mathbf{K})$ is called the **Galois group** of the extension and written $\mathrm{Gal}(\mathbf{K}/\mathbf{k})$. The first result of Galois theory, proved in every relevant book, is

(4.3.1) THE GALOIS CORRESPONDENCE. *Let $\mathbf{K}/\mathbf{k}$ be a Galois extension with group G. Then there is an inclusion-reversing bijective correspondence between intermediate fields $\mathbf{k} \subset \mathbf{F} \subset \mathbf{K}$ and subgroups $H \subset G$, as follows.*

$$\{\textit{intermediate fields}\} \longrightarrow \{\textit{subgroups}\} \qquad \textit{by} \qquad \mathbf{F} \mapsto G_{\mathbf{F}}$$

where $G_{\mathbf{F}} = \{\sigma \in G : x^\sigma = x \text{ for all } x \in \mathbf{F}\}$ is the subgroup of G that fixes $\mathbf{F}$ pointwise, and

$$\{\textit{subgroups}\} \longrightarrow \{\textit{intermediate fields}\} \qquad \textit{by} \qquad H \mapsto \mathbf{K}^H$$

where $\mathbf{K}^H = \{x \in \mathbf{K} : x^\sigma = x \text{ for all } \sigma \in H\}$ is the subfield of $\mathbf{K}$ fixed pointwise by H. In particular, only $\mathbf{k}$ is fixed pointwise by G.

The relations

$$|G_{\mathbf{F}}| = [\mathbf{K} : \mathbf{F}] \qquad \textit{and} \qquad [\mathbf{K}^H : \mathbf{k}] = [G : H]$$

hold for all intermediate fields $\mathbf{F}$ and subgroups H. In particular, $|G| = [\mathbf{K} : \mathbf{k}]$. For any intermediate field $\mathbf{F}$, the extension $\mathbf{K}/\mathbf{F}$ is Galois with group $G_{\mathbf{F}}$, while the extension $\mathbf{F}/\mathbf{k}$ is Galois if and only if $H \lhd G$, in which case $\mathrm{Gal}(\mathbf{F}/\mathbf{k}) \cong G/H$.

For an example of all this, Exercise 4.2.8 shows that the fifth root of unity $\zeta_5 = e^{2\pi i/5} \in \mathbf{C}$ has minimal polynomial $m = 1 + T + T^2 + T^3 + T^4$, and $\mathrm{spl}_{\mathbf{Q}}(m) = \mathbf{Q}(\zeta_5)$ has degree 4 over $\mathbf{Q}$. The extension $\mathbf{Q}(\zeta_5)/\mathbf{Q}$ is Galois. Let G denote its group, of order 4. Any $\sigma \in G$ is specified by its action on ζ_5 and must take ζ_5 to another root ζ_5^j of m. In particular, the automorphism $\sigma : \zeta_5 \mapsto \zeta_5^2$ has order 4 since $\sigma^2 = \sigma \circ \sigma : \zeta_5 \overset{\sigma}{\mapsto} \zeta_5^2 = \zeta_5\zeta_5 \overset{\sigma}{\mapsto} \zeta_5^2\zeta_5^2 = \zeta_5^{-1}$ is not the identity. Thus $G = C_4$ and its only nontrivial proper subgroup is C_2 generated by σ^2. Corresponding to the chain of groups $\{1\} \lhd C_2 \lhd G$ is the tower of Galois extensions $\mathbf{Q} \subset \mathbf{F} \subset \mathbf{Q}(\zeta_5)$, where $\mathbf{F} = \mathbf{Q}(\zeta_5)^{\langle\sigma^2\rangle}$ is the only proper intermediate field, of degree 2 over $\mathbf{Q}$. (See Figure 4.3.1.) Explicitly $\mathbf{F} = \mathbf{Q}(\zeta_5 + \zeta_5^{-1})$ since $\zeta_5 + \zeta_5^{-1}$ is fixed by σ^2 but not by σ. Thus $\zeta_5 + \zeta_5^{-1}$ must satisfy a quadratic polynomial over $\mathbf{Q}$. Indeed, from $m(\zeta_5) = 0$ compute

$$0 = \zeta_5^2 + \zeta_5^{-2} + \zeta_5 + \zeta_5^{-1} + 1 = (\zeta_5 + \zeta_5^{-1})^2 + (\zeta_5 + \zeta_5^{-1}) - 1,$$

so that $(\zeta_5 + \zeta_5^{-1})^2 = 1 - (\zeta_5 + \zeta_5^{-1})$. Since $\zeta_5 + \zeta_5^{-1}$ is positive, this polynomial relation identifies it, perhaps surprisingly, as the golden ratio g from Chapter 2.

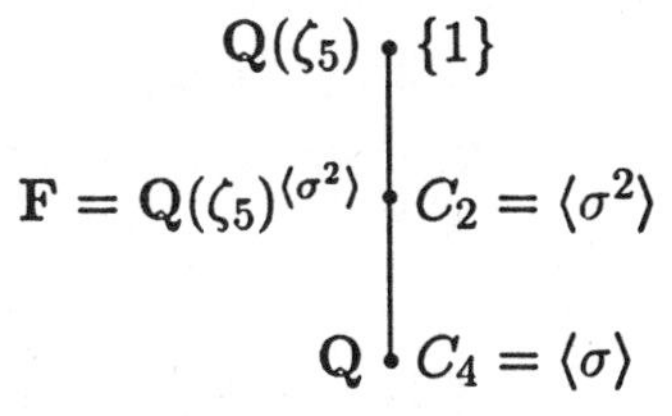

$\mathbf{Q}(\zeta_5)$ — $\{1\}$

$\mathbf{F} = \mathbf{Q}(\zeta_5)^{\langle\sigma^2\rangle}$ — $C_2 = \langle\sigma^2\rangle$

$\mathbf{Q}$ — $C_4 = \langle\sigma\rangle$

Figure 4.3.1. The Galois Correspondence for $\mathbf{Q}(\zeta_5)/\mathbf{Q}$

For another example, the Galois extension $\mathbf{Q}(\sqrt[3]{2}, \zeta_3)/\mathbf{Q}$ has group $G = \langle\sigma, \tau\rangle$ where $(\sqrt[3]{2})^\sigma = \zeta_3\sqrt[3]{2}$, $(\zeta_3)^\sigma = \zeta_3$, and $(\sqrt[3]{2})^\tau = \sqrt[3]{2}$, $(\zeta_3)^\tau = \zeta_3^2$. These satisfy $\sigma^3 = \tau^2 = 1$ and $\sigma\tau = \tau\sigma^{-1}$, showing that $G = D_3$. The subgroups and intermediate fields are shown in Figure 4.3.2, where composition is left-to-right as usual and the double lines indicate Galois extensions.

Here is a nice connection between Galois groups and polynomial factorization.

(4.3.2) PROPOSITION. *Let $\mathbf{k}$ be a field, let $h \in \mathbf{k}[T]$ be a monic separable polynomial, let $\mathbf{K} = \mathrm{spl}_{\mathbf{k}}(h)$, so $\mathbf{K}/\mathbf{k}$ is Galois, and let $G = \mathrm{Gal}(\mathbf{K}/\mathbf{k})$. Then h is irreducible in $\mathbf{k}[T]$ if and only if G acts transitively on the roots of h in $\mathbf{K}$. More generally, the irreducible factors of h in $\mathbf{k}[T]$ are $h_i = \prod_{r \in \mathcal{O}_i}(T - r)$ where the $\mathcal{O}_i$ are the G-orbits of roots of h in $\mathbf{K}$.*

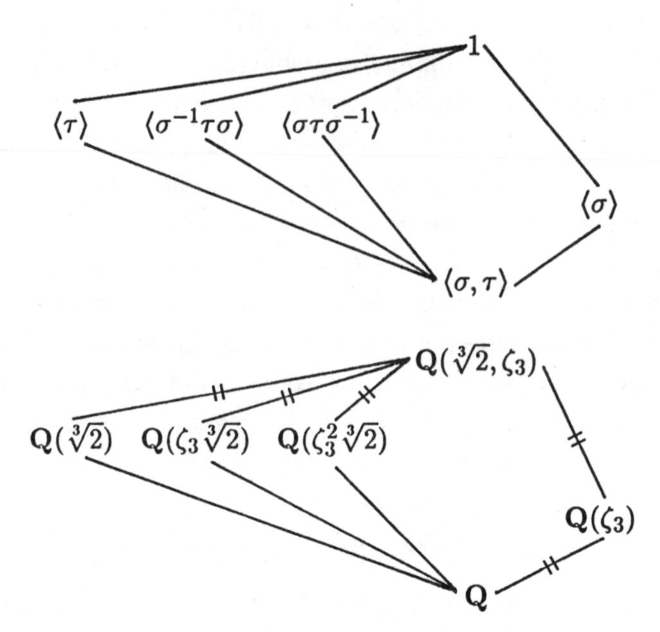

Figure 4.3.2. The Galois Correspondence for $\mathbf{Q}(\sqrt[3]{2},\zeta_3)/\mathbf{Q}$

PROOF. Certainly $h = \prod_i h_i$. Any polynomial in $\mathbf{K}[T]$ lies in $\mathbf{k}[T]$ if and only if it is fixed by G. Thus each h_i, being fixed by G, lies in $\mathbf{k}[T]$. On the other hand, no factor of an h_i is fixed by G, so each h_i is irreducible in $\mathbf{k}[T]$. □

Exercises

4.3.1. Let $\mathbf{k}$ be a field, R be a ring, and $\varphi : \mathbf{k} \longrightarrow R$ be a homomorphism. Show that unless φ is identically zero, φ is injective. In particular, a surjective field homomorphism must be an isomorphism.

4.3.2. Show that any splitting field extension is normal.

4.3.3. Show that the extension $\mathbf{Q}(\sqrt[3]{2})/\mathbf{Q}$, where $\sqrt[3]{2}$ is the real cube root of 2, is not normal.

4.3.4. This exercise shows that any algebraic extension $\mathbf{K}/\mathbf{k}$, where $\mathbf{k}$ is a finite field of characteristic p, is separable. It suffices to show that any irreducible polynomial $m \in \mathbf{k}[T]$ is separable, i.e., shares no factor with its derivative. As argued in the text for characteristic zero, this can only fail if $m' = 0$. In characteristic p, $m' = 0$ means that $m(T) = \tilde{m}(T^p)$ for some $\tilde{m} \in \mathbf{k}[T]$. Use Exercise 4.1.4 and finiteness of $\mathbf{k}$ to show that the p-power map $x \mapsto x^p$ is an automorphism (self-isomorphism) of $\mathbf{k}$. It follows that $\tilde{m}(T^p) = n(T)^p$ for some $n \in \mathbf{k}[T]$, contradicting irreducibility of m.

4.3.5. For an example of an inseparable extension, let p be a prime, let $\mathbf{k} = \mathbf{F}_p(W)$ where W is transcendental over $\mathbf{F}_p$, and let $\mathbf{K}$ be the splitting field over $\mathbf{k}$ of the polynomial $f = T^p - W \in \mathbf{k}[T]$. If $Z \in \mathbf{K}$ is a root of f, so that $W = Z^p$, then in fact $f = (T - Z)^p$ by Exercise 4.1.4, and $\mathbf{K} = \mathbf{k}(Z)$. Show that $Z \notin \mathbf{k}$ by supposing a relation $Z = g(W)/h(W)$ and taking pth powers. Thus the minimal polynomial of Z over $\mathbf{k}$ is $m = (T - Z)^e$ for some $e > 1$, and the extension $\mathbf{K}/\mathbf{k}$ is inseparable. In fact, $e = p$, meaning that f is irreducible over $\mathbf{k}$. To see this, note that Z^e lies in $\mathbf{k}$ since $m \in \mathbf{k}[T]$. Since also $Z^p \in \mathbf{k}$, it follows that $Z^{\gcd(e,p)} \in \mathbf{k}$ since $\gcd(e,p)$ takes the form $ae + bp$ for integers a and b. Since we have already shown that $Z \notin \mathbf{k}$, it follows that $\gcd(e,p) > 1$, i.e., $e = p$.

4.3.6. Describe the Galois Correspondence for the extension $\mathbf{Q}(\sqrt{2}, \sqrt{3})/\mathbf{Q}$.

4. The rotation group extension

Our concern is with a special case of all these ideas. To study the extension $\mathbf{C}(Z)/\mathbf{C}(f_\Gamma)$, where Z is a formal symbol and f_Γ is a formal rational expression in Z, introduce a new symbol

$$W = f_\Gamma(Z).$$

Like Z, W is transcendental over $\mathbf{C}$, though Z and W are algebraically related by the definition of W. Letting $f_\Gamma(Z) = g(Z)/h(Z)$, an equivalent defining relation between Z and W is $g(Z) - Wh(Z) = 0$, or

$$p_W(Z) = 0 \qquad \text{where} \qquad p_W(T) = g(T) - Wh(T) \in \mathbf{C}(W)[T].$$

Thus we are studying the extension $\mathbf{C}(Z)/\mathbf{C}(W)$ obtained by adjoining the root Z of p_W to $\mathbf{C}(W)$. In fact $\mathbf{C}(Z)$ is the splitting field $\mathrm{spl}_{\mathbf{C}(W)}(p_W)$ since the roots of p_W are $\Gamma Z = \{\gamma Z : \gamma \in \Gamma\}$, all of which are rational in Z and thus elements of $\mathbf{C}(Z)$. As a splitting field extension in characteristic zero, $\mathbf{C}(Z)/\mathbf{C}(W)$ is Galois. Let G denote its Galois group. A close connection between G and Γ seems natural, but expressing it precisely requires some care. The problem is that the elements of Γ are fractional linear transformations acting on $\widehat{\mathbf{C}}$, while G consists of automorphisms of $\mathbf{C}(Z)$.

Any fractional linear transformation $\gamma \in \mathrm{PSL}_2(\mathbf{C})$ defines—but is not quite the same as—a corresponding automorphism $\gamma^* \in \mathrm{Aut}_{\mathbf{C}}(\mathbf{C}(Z))$. The automorphism, called the **pullback** of γ, is given by composition with γ; in symbols,

$$\gamma^*(r) = r \circ \gamma \text{ (i.e., } \gamma^*(r(Z)) = r(\gamma(Z))) \quad \text{for all } r \in \mathbf{C}(Z).$$

(Exercise 4.4.1 asks why γ^* lies in $\mathrm{Aut}_{\mathbf{C}}(\mathbf{C}(Z))$.) Thus the pullback is a map

$$^* : \mathrm{PSL}_2(\mathbf{C}) \longrightarrow \mathrm{Aut}_{\mathbf{C}}(\mathbf{C}(Z)).$$

Exercise 4.4.3 shows that this map is a bijection.

Now we can return to the connection between the groups Γ and $G = \mathrm{Gal}(\mathbf{C}(Z)/\mathbf{C}(W))$. Since $f_\Gamma(\gamma Z) = W$ for all $\gamma \in \Gamma$, the pullback identifies Γ with a subset of G. Conversely, any automorphism $\sigma \in G$ is determined by its action on Z and must take Z to another root of p_W, so it must act on Z as some $\gamma \in \Gamma$. Thus $\sigma = \gamma^*$ and the pullback identifies Γ with G. Use this identification from now on and view G as Γ acting on rational functions from the right via composition. Writing this action as r^γ is compatible with the superscript notation for automorphisms, meaning

$$r^{\gamma_1\gamma_2} = r \circ (\gamma_1\gamma_2) = (r \circ \gamma_1) \circ \gamma_2 = (r^{\gamma_1})^{\gamma_2}.$$

Since now $\mathrm{Gal}(\mathbf{C}(Z)/\mathbf{C}(W)) = \Gamma$ as sets, Galois theory says that $[\mathbf{C}(Z) : \mathbf{C}(W)] = |\Gamma|$. This last equality can also be proved without Galois theory (Exercise 4.4.4 and again in Exercise 4.4.5 by Gauss' Lemma from Exercise 4.4.2).

Even though the pullback naturally identifies the sets $\mathrm{PSL}_2(\mathbf{C})$ and $\mathrm{Aut}_{\mathbf{C}}(\mathbf{C}(Z))$, it is not a group isomorphism. For any $\gamma_1, \gamma_2 \in \mathrm{PSL}_2(\mathbf{C})$ and $r \in \mathbf{C}(Z)$,

$$(\gamma_1 \circ \gamma_2)^* r = r \circ (\gamma_1 \circ \gamma_2) = (r \circ \gamma_1) \circ \gamma_2 = \gamma_2^*(\gamma_1^*(r)) = (\gamma_2^* \circ \gamma_1^*) r,$$

meaning $(\gamma_1 \circ \gamma_2)^* = \gamma_2^* \circ \gamma_1^*$. Thus the pullback reverses products; it is an **anti-homomorphism** or a **contravariant** map. The clever trick that modifies the pullback into a homomorphism is combining it with another product-reversing operation, the inversion map on $\mathrm{PSL}_2(\mathbf{C})$. Define the **inverse-pullback**

$$^{-*} : \mathrm{PSL}_2(\mathbf{C}) \longrightarrow \mathrm{Aut}_{\mathbf{C}}(\mathbf{C}(Z))$$

by $\gamma^{-*}(r) = r \circ \gamma^{-1}$ for all $\gamma \in \mathrm{PSL}_2(\mathbf{C})$ and $r \in \mathbf{C}(Z)$. This is a group isomorphism (Exercise 4.4.6). Thus $\mathrm{Gal}(\mathbf{C}(Z)/\mathbf{C}(W))$ is not only elementwise identified with Γ, the two groups are isomorphic; but the isomorphism is not the natural identification.

Exercises

4.4.1. Show that for any $\gamma \in \mathrm{PSL}_2(\mathbf{C})$, the pullback γ^* is an automorphism of $\mathbf{C}(Z)$ over $\mathbf{C}$.

4.4.2. (a) A polynomial $p \in \mathbf{Z}[T]$ is called **primitive** if the greatest common divisor of its coefficients is 1. Show that the product of primitive polynomials is again primitive.

(b) Use (a) to show that if the polynomial $p \in \mathbf{Z}[T]$ factors nontrivially in $\mathbf{Q}[T]$ then it factors nontrivially in $\mathbf{Z}[T]$. (Hint: first show that any nonzero polynomial $q \in \mathbf{Q}[T]$ takes the form $q = r_q\tilde{q}$ where r_q is a rational number and $\tilde{q}$ is primitive in $\mathbf{Z}[T]$. Apply this to the factors of p.)

(c) Confirm that your proof of (b) goes through verbatim when $\mathbf{Z}$ and $\mathbf{Q}$ are replaced respectively by any unique factorization domain and its field of quotients. The results in (a), (b) and (c) all go under the name **Gauss' Lemma**.

4.4.3. This exercise shows that the pullback $^* : \mathrm{PSL}_2(\mathbf{C}) \longrightarrow \mathrm{Aut}_{\mathbf{C}}(\mathbf{C}(Z))$ is a bijection.

(a) Show that the pullback is an injection.

(b) To show that the pullback surjects, take $\sigma \in \mathrm{Aut}_{\mathbf{C}}(\mathbf{C}(Z))$. Then in particular $\sigma(Z) = g(Z)/h(Z) \stackrel{\text{call}}{=} W' \in \mathbf{C}(Z)$, where $g, h \in \mathbf{C}[T]$ share no factor and aren't both constant, and this completely specifies σ. Show that $[\mathbf{C}(Z) : \mathbf{C}(W')] = \max\{\deg(g), \deg(h)\}$ by using the ideas of the section: let $p(T) = g(T) - W'h(T) \in \mathbf{C}(W')[T]$; then p is satisfied by Z and is irreducible in $\mathbf{C}(W')[T]$ by Gauss' Lemma (Exercise 4.4.2), so it is the minimal polynomial of Z over $\mathbf{C}(W')$, up to scalar multiple. Since σ is an automorphism, $\mathbf{C}(W') = \mathbf{C}(Z)$, i.e., $\max\{\deg(g), \deg(h)\} = 1$, i.e., σ acts on Z as an element γ of $\mathrm{PSL}_2(\mathbf{C})$, i.e., $\sigma = \gamma^*$.

4.4.4. Prove without Galois theory that $[\mathbf{C}(Z) : \mathbf{C}(W)] = |\Gamma|$ by justifying the following argument as necessary. The group Γ fixes $\mathbf{C}(W)$ and acts transitively on the roots ΓZ of p_W. The elements of ΓZ occur once each as roots since they are distinct and $\deg(p_W) = |\Gamma|$. No proper factor of p_W is fixed by Γ, hence no proper factor lies in $\mathbf{C}(W)[T]$, so p_W is irreducible over $\mathbf{C}(W)$ and $[\mathbf{C}(Z) : \mathbf{C}(W)] = \deg(p_W) = |\Gamma|$. This proof that p_W is irreducible is similar to the proof of Proposition 4.3.2, which relies on unproven facts about the Galois Correspondence. Why doesn't the argument here rely on anything we haven't proved?

4.4.5. Use Exercise 4.4.2(c) to show that p_W from the section is irreducible by considering it first as an element of $\mathbf{C}[W][T] = \mathbf{C}[T][W]$. (The ring $\mathbf{k}[T_1, \ldots, T_n]$ of polynomials over any field in any finite number of variables is a unique factorization domain—see, for example, [Co-Li-O'S].) As in the preceding exercise, the relation $[\mathbf{C}(Z) : \mathbf{C}(W)] = |\Gamma|$ follows.

4.4.6. Show that the inverse-pullback $^{-*} : \mathrm{PSL}_2(\mathbf{C}) \longrightarrow \mathrm{Aut}_{\mathbf{C}}(\mathbf{C}(Z))$ is a group isomorphism. (Hint: the inverse-pullback is the composition of the inversion map on $\mathrm{PSL}_2(\mathbf{C})$ and the pullback, so it suffices to show that each of these maps is contravariant and bijective.)

5. The Radical Criterion

Let $p \in \mathbf{k}[T]$ be a polynomial over a field. This section develops the criterion that Galois gave to determine whether the roots of p can be expressed in radicals over $\mathbf{k}$. In field language, the question is whether the splitting field $\mathrm{spl}_{\mathbf{k}}(p)$ is contained in a field obtained by adjoining radicals, starting from the base field $\mathbf{k}$. Thus the following definition is natural.

(4.5.1) DEFINITION. *Let* $\mathbf{k}$ *be a field. A* **root tower over k** *is a sequence of fields* $\mathbf{k} \subset \mathbf{k}_1 \subset \mathbf{k}_2 \subset \cdots \subset \mathbf{k}_d$ *such that*

$$
\begin{aligned}
\mathbf{k}_1 &= \mathbf{k}(a_1) && \text{where } a_1^{n_1} \in \mathbf{k} \text{ for some } n_1 \in \mathbf{Z}^+,\\
\mathbf{k}_2 &= \mathbf{k}_1(a_2) && \text{where } a_2^{n_2} \in \mathbf{k}_1 \text{ for some } n_2 \in \mathbf{Z}^+,\\
&\vdots && \vdots\\
\mathbf{k}_d &= \mathbf{k}_{d-1}(a_d) && \text{where } a_d^{n_d} \in \mathbf{k}_{d-1} \text{ for some } n_d \in \mathbf{Z}^+.
\end{aligned}
$$

The exponents $n_1, \ldots, n_d$ *are taken to be minimal and are greater than 1, of course.*

An extension $\mathbf{K}/\mathbf{k}$ *is* **constructible by radicals** *if there there exists a root tower* $\mathbf{k} \subset \cdots \subset \mathbf{k}_d$ *with* $\mathbf{K} \subset \mathbf{k}_d$.

Galois showed that solvability of the polynomial p by radicals is equivalent to a condition on the group $\mathrm{Gal}(\mathrm{spl}_{\mathbf{k}}(p)/\mathbf{k})$. Since finite groups are comparatively simple objects, this is a tremendous reduction. The relevant definition is

(4.5.2) DEFINITION. *A finite group* Γ *is* **solvable** *if there exists a chain of subgroups*

$$\{1\} = \Gamma_d \triangleleft \Gamma_{d-1} \triangleleft \cdots \triangleleft \Gamma_1 \triangleleft \Gamma_0 = \Gamma$$

each normal in the next and with each quotient Γ_i/Γ_{i+1} *a cyclic group* C_{n_i}. *This chain is called a* **subnormal series** *for* Γ.

This definition is unaffected if the quotients are stipulated to be abelian rather than cyclic (Exercise 4.5.1). Any subgroup or quotient group of a solvable group is again solvable (Exercise 4.5.2).

Galois' criterion in its most general form is more than we need. We are working over the field $\mathbf{C}$, which contains all roots of unity and has characteristic zero. In a root tower whose base field contains enough roots of unity, each extension $\mathbf{k}_i/\mathbf{k}_{i-1}$ is **cyclic**, meaning it is a Galois extension with cyclic group, in this case C_{n_i} (Exercise 4.5.3). Roots of unity will also play a role in Lagrange's Lemma, to be stated soon. In characteristic zero, any finite extension $\mathbf{K}/\mathbf{k}$ is separable and $\mathbf{K}$ has a **Galois closure over** $\mathbf{k}$, meaning the smallest superfield $\mathbf{K}'$ of $\mathbf{K}$ such that $\mathbf{K}'/\mathbf{k}$ is Galois. This is the field generated by all $\mathbf{k}$-embedded images of $\mathbf{K}$ into the algebraic closure of $\mathbf{k}$—we skip the details.

(4.5.3) RADICAL CRITERION (OVER $\mathbf{C}$). *Let* $\mathbf{k}$ *be a field containing* $\mathbf{C}$, *let* $\mathbf{K}/\mathbf{k}$ *be a finite extension, and let* $\mathbf{K}'$ *be the Galois closure of* $\mathbf{K}$ *over* $\mathbf{k}$. *Then*

$$\mathbf{K}/\mathbf{k} \text{ is constructible by radicals} \quad \Longleftrightarrow \quad \mathrm{Gal}(\mathbf{K}'/\mathbf{k}) \text{ is solvable.}$$

When these conditions hold, there exists a root tower over $\mathbf{k}$ *whose top field is* $\mathbf{K}'$. *In particular, if* $\mathbf{K}/\mathbf{k}$ *is Galois with a solvable group then* $\mathbf{K}$ *itself is the top field in a root tower over* $\mathbf{k}$.

PROOF. ($\Longrightarrow$) Suppose $\mathbf{K}/\mathbf{k}$ is constructible by radicals. Applying the Galois Correspondence to a root tower $\mathbf{k} \subset \cdots \subset \mathbf{k}_d$ gives Figure 4.5.1. The extension $\mathbf{k}_1/\mathbf{k}$ is Galois with a cyclic group, say C_{n_1}, so $\Gamma_1 \triangleleft \Gamma_0$ and $\Gamma_0/\Gamma_1 = C_{n_1}$. Proceeding up the tower in this fashion shows that $\mathrm{Gal}(\mathbf{k}_d/\mathbf{k})$ is solvable. Recall that $\mathbf{K} \subset \mathbf{k}_d$; the Galois closure $\mathbf{K}'$ also lies in $\mathbf{k}_d$ (Exercise 4.5.4), so $\mathrm{Gal}(\mathbf{K}'/\mathbf{k})$ is a quotient of $\mathrm{Gal}(\mathbf{k}_d/\mathbf{k})$ and is therefore solvable as well. □

Proving the other direction requires some preliminary results.

(4.5.4) LINEAR INDEPENDENCE OF AUTOMORPHISMS. *Let* $\mathbf{k}$ *be a field. Any set* $\mathcal{S}$ *of automorphisms of* $\mathbf{k}$ *is linearly independent, meaning that there is no relation*

$$a_1\sigma_1 + \cdots + a_n\sigma_n = 0 \tag{4.5.5}$$

with nonzero $a_1, \ldots, a_n \in \mathbf{k}$ *and distinct* $\sigma_1, \ldots, \sigma_n \in \mathcal{S}$.

$$\begin{array}{r|l}
\mathbf{k}_d & \Gamma_d = \{1\} \\
\mathbf{k}_{d-1} & \Gamma_{d-1} \\
\vdots & \vdots \\
\mathbf{k}_1 & \Gamma_1 \\
\mathbf{k} & \Gamma_0 = \mathrm{Gal}(\mathbf{k}_d/\mathbf{k})
\end{array}$$

Figure 4.5.1. Root tower and corresponding chain of subgroups

The proof is Exercise 4.5.5.

(4.5.6) LAGRANGE LEMMA. *Let n be a positive integer, and let $\mathbf{k}$ be a field containing a primitive nth root of unity ζ_n. Let $\mathbf{K}/\mathbf{k}$ be cyclic of degree n, i.e., Galois with group C_n. Then there exists an element $a \in \mathbf{K}$ such that $\mathbf{k}(a) = \mathbf{K}$ and $a^n \in \mathbf{k}$. That is, $\mathbf{K}$ is generated over $\mathbf{k}$ by an nth root.*

Note that n is the smallest positive exponent such that $a^n \in \mathbf{k}$, otherwise a would satisfy a polynomial of smaller degree over $\mathbf{k}$, violating $[\mathbf{k}(a) : \mathbf{k}] = [\mathbf{K} : \mathbf{k}] = |\mathrm{Gal}(\mathbf{K}/\mathbf{k})| = |C_n| = n$. The ζ_n in the Lagrange Lemma need not be the familiar complex number $\zeta_n = e^{2\pi i/n}$, for example in the case of finite fields.

PROOF. Let σ generate the Galois group. Define $\tau = \sum_{i=0}^{n-1} \zeta_n^i \sigma^i$. By linear independence, there exists $b \in \mathbf{K}$ such that $a = b^\tau$ is nonzero. Compute

$$a^\sigma = (b^\tau)^\sigma = \Big(\sum_{i=0}^{n-1} \zeta_n^i b^{\sigma^i}\Big)^\sigma = \sum_{i=0}^{n-1} \zeta_n^i b^{\sigma^{i+1}} = \zeta_n^{-1} b^\tau = \zeta_n^{-1} a.$$

Thus $a^{\sigma^i} = \zeta_n^{-i} a \neq a$ for $i = 1, \ldots, n-1$, showing that the fixing subgroup of $\mathbf{k}(a)$ is $\{1\}$, so by the Galois Correspondence $\mathbf{k}(a) = \mathbf{K}$. And since $(a^n)^\sigma = (a^\sigma)^n = (\zeta^{-1}a)^n = a^n$, it follows that a^n is fixed by the entire Galois group, so $a^n \in \mathbf{k}$. □

Now we can prove the other direction of the Radical Criterion.

PROOF. ($\Longleftarrow$) Suppose $\mathrm{Gal}(\mathbf{K}'/\mathbf{k})$ is solvable. The Galois Correspondence gives Figure 4.5.2. Since $\Gamma_1 \triangleleft \Gamma_0$ and $\Gamma_0/\Gamma_1 = C_{n_1}$, the Lagrange Lemma shows that $\mathbf{k}_1 = \mathbf{k}(a_1)$ with $a_1^{n_1} \in \mathbf{k}$. Proceeding down the subnormal series

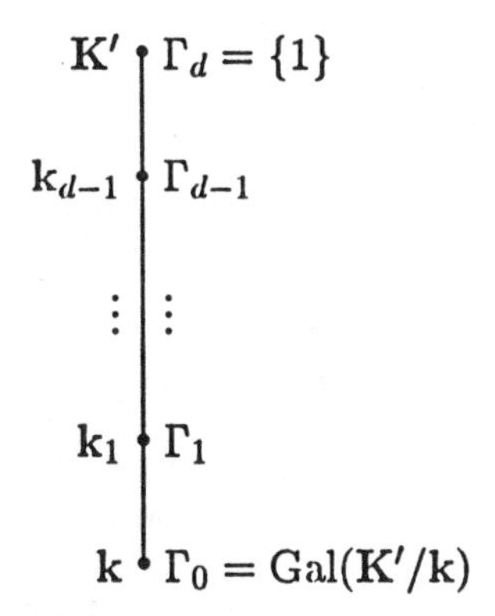

Figure 4.5.2. Subnormal series and corresponding tower of fields

shows that the intermediate fields form a root tower. Since $\mathbf{K} \subset \mathbf{K}'$, $\mathbf{K}/\mathbf{k}$ is constructible by radicals. □

Proving the last two statements of the Radical Criterion is Exercise 4.5.6.

Exercises

4.5.1. Show that changing the word "cyclic" to "abelian" in Definition 4.5.2 does not affect which groups are solvable.

4.5.2. (a) Prove the Third Isomorphism Theorem of group theory: Let G be a group with normal subgroups $N \subset H$. Then $H/N \triangleleft G/N$ and there is a natural isomorphism $(G/N)/(H/N) \xrightarrow{\sim} G/H$. (Show that the map $G/N \longrightarrow G/H$ given by $gN \mapsto gH$ is well-defined, is surjective, and has kernel H/N.)

(b) Show that any subgroup or quotient group of a solvable group Γ is again solvable. (If H is a subgroup, intersect a subnormal series for Γ termwise with H. Show that each term in the resulting chain is a normal subgroup of the next. To study the quotients, note that the map $\Gamma_i \cap H \longrightarrow \Gamma_i \longrightarrow \Gamma_i/\Gamma_{i+1}$ has kernel $\Gamma_{i+1} \cap H$ and cite the First Isomorphism Theorem. Similarly, if Γ/N is a quotient group, map the subnormal series for Γ termwise to Γ/N. Again each term in the resulting chain is a normal subgroup of the next. Use the First Isomorphism Theorem to show that the terms are isomorphic to $\Gamma_i/(\Gamma_i \cap N)$, and then use the Second and Third Isomorphism Theorems to study the new quotients.)

4.5.3. Show that in a root tower over a base field $\mathbf{k}$, each extension $\mathbf{k}_i/\mathbf{k}_{i-1}$ is Galois with group C_{n_i} provided that $\mathbf{k}$ contains a primitive n_ith root of unity, i.e., ζ_{n_i} such that $\zeta_{n_i}^{n_i} = 1 \neq \zeta_{n_i}^j$ for $1 \leq j < n_i$.

4.5.4. In the proof of ($\Longrightarrow$) of the Radical Criterion , explain why $\mathbf{K}' \subset \mathbf{k}_d$.

4.5.5. Prove the linear independence of automorphisms as follows. Assume a relation $\sum_{i=1}^{n} a_i\sigma_i = 0$ of the form (4.5.5) with minimal n. Then $n \geq 2$. There exists $x \in \mathbf{k}$ such that $x^{\sigma_1} \neq x^{\sigma_2}$. Show that $\sum_{i=1}^{n} a_i x^{\sigma_1}\sigma_i = 0$ and $\sum_{i=1}^{n} a_i x^{\sigma_i}\sigma_i = 0$. Subtracting these relations gives a contradiction.

4.5.6. Complete the proof of the Radical Criterion.

6. Algebraic inversion of the nonicosahedral invariants

Each finite rotation group Γ except the icosahedral group Γ_I is solvable. Each nonicosahedral extension $\mathbf{C}(Z)/\mathbf{C}(W)$, being Galois, can therefore be constructed by a sequence of root adjunctions, cf. the last sentence of the Radical Criterion. On the other hand, the icosahedral group is not solvable (Exercise 4.6.1), so no sequence of root adjunctions to $\mathbf{C}(W)$ will produce a superfield of $\mathbf{C}(Z)$ in the icosahedral case.

To invert the cyclic and dihedral equations consider the chain of subgroups $\{1\} \triangleleft C_n \triangleleft D_n$ with corresponding tower of cyclic extensions shown in Figure 4.6.1. The field containments show that f_{C_n} is rational in Z, and f_{D_n} is rational in f_{C_n}, as we already know from the relations $f_{C_n} = Z^n$ and $f_{D_n} = (f_{C_n}+1)^2/(4f_{C_n})$. These two relations can be recast as polynomial conditions. Indeed, setting $W = f_{D_n}$ in the second relation shows that f_{C_n} satisfies the polynomial

$$m_{f_{C_n}/f_{D_n}}(T;W) = T^2 + (2-4W)T + 1$$

whose solutions involve a square root over $\mathbf{C}(W)$. (The notation $m(T;W)$ is meant to suggest a polynomial in T whose coefficients depend on W.)

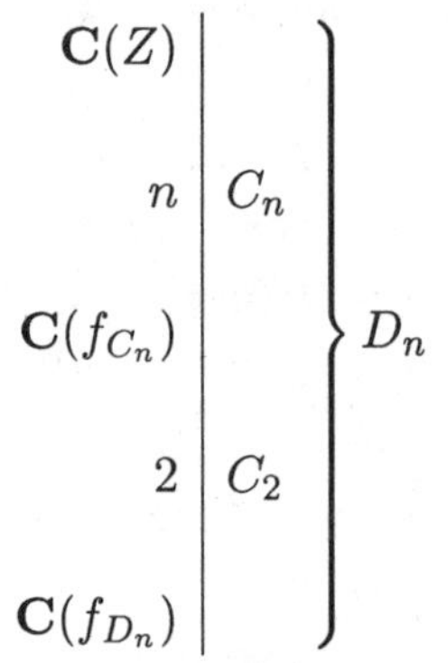

$$\begin{array}{r|ll} \mathbf{C}(Z) & & \\ n & C_n & \\ \mathbf{C}(f_{C_n}) & & D_n \\ 2 & C_2 & \\ \mathbf{C}(f_{D_n}) & & \end{array}$$

Figure 4.6.1. Tower of fields for $\{1\} \triangleleft C_n \triangleleft D_n$

Similarly, setting $W = f_{C_n}$ in the first relation shows that Z satisfies the polynomial

$$m_{Z/f_{C_n}}(T;W) = T^n - W$$

whose solutions involve an nth root $\sqrt[n]{W}$ over $\mathbf{C}(W)$. The dihedral inversion now reduces to a two-step procedure:

(4.6.1) ALGORITHM. *To invert any specific equation* $f_{D_n}(z) = w$ *for* $w \in \mathbf{C}$,

1. *Find a root* w' *of* $m_{f_{C_n}/f_{D_n}}(T;w)$.
2. *Find a root* z_0 *of* $m_{Z/f_{C_n}}(T;w')$. *The* z*-values over* w *are then* $D_n(z_0)$.

Skipping the first step and taking $C_n(z_0)$ *in the second inverts the equation* $f_{C_n}(z) = w'$.

To invert the tetrahedral and octahedral equations, take the chain of subgroups $\{1\} \triangleleft C_2 \triangleleft D_2 \triangleleft \Gamma_T \triangleleft \Gamma_O$ (the D_2 is the Klein four-group V) with corresponding tower of cyclic extensions shown in Figure 4.6.2. The bottom field containment shows that f_O is a rational expression in f_T. To find this

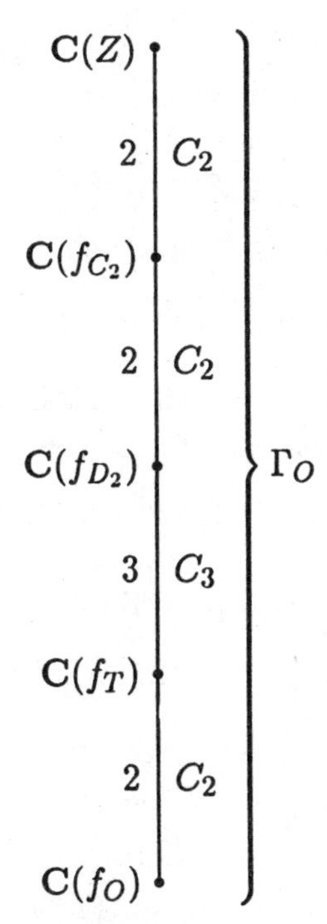

Figure 4.6.2. Tower of fields for $\{1\} \triangleleft C_2 \triangleleft D_2 \triangleleft \Gamma_T \triangleleft \Gamma_O$

expression, recall the relations

$$f_T = \frac{(F_{2,T}^3)_*}{(F_{1,T}^3)_*}, \qquad f_O = \frac{(F_{2,O}^3)_*}{108(F_{1,O}^4)_*}, \qquad F_{3,T}^2 = \frac{F_{2,T}^3 - F_{1,T}^3}{12i\sqrt{3}}.$$

Geometry shows that $F_{1,O} = F_{3,T}$ and $F_{2,O} = F_{1,T}F_{2,T}$ up to constant multiples, and in fact the forms are normalized so that these identities hold. Thus

$$\begin{aligned} f_O &= \frac{(F_{2,O}^3)_*}{108(F_{1,O}^4)_*} = \frac{(F_{1,T}^3 F_{2,T}^3)_*}{108(F_{3,T}^4)_*} = \frac{(12i\sqrt{3})^2(F_{1,T}^3 F_{2,T}^3)_*}{108((F_{2,T}^3)_* - (F_{1,T}^3)_*)^2} \\ &= \frac{-4(F_{1,T}^3 F_{2,T}^3)_*}{(F_{1,T}^3)_*^2(f_T - 1)^2} = \frac{-4f_T}{(f_T - 1)^2}. \end{aligned}$$

Setting $W = f_O$ in this rational relation shows that f_T satisfies the polynomial

$$m_{f_T/f_O}(T; W) = T^2 + (4/W - 2)T + 1$$

whose solutions involve a square root over $\mathbf{C}(W)$.

Similarly, f_T must be rational in f_{D_2}. Recall that

$$f_{D_2} = \frac{(Z^2+1)^2}{4Z^2}, \qquad f_T = \left(\frac{Z^4 + 2i\sqrt{3}Z^2 + 1}{Z^4 - 2i\sqrt{3}Z^2 + 1}\right)^3,$$

and compute $f_{D_2} + \zeta_3 = \dfrac{Z^4 + 2Z^2 + 1}{4Z^2} - \dfrac{2Z^2}{4Z^2} + \dfrac{2i\sqrt{3}Z^2}{4Z^2} = \dfrac{Z^4 + 2i\sqrt{3}Z^2 + 1}{4Z^2}$.

Likewise, $f_{D_2} + \zeta_3^2 = \dfrac{Z^4 - 2i\sqrt{3}Z^2 + 1}{4Z^2}$. So $f_T = \left(\dfrac{f_{D_2} + \zeta_3}{f_{D_2} + \zeta_3^2}\right)^3$, and setting $W = f_T$ in this rational relation shows that f_{D_2} satisfies the polynomial

$$m_{f_{D_2}/f_T}(T; W) = (W-1)T^3 + 3(W\zeta_3^2 - \zeta_3)T^2 + 3(W\zeta_3 - \zeta_3^2)T + (W-1).$$

Since $m_{f_{D_2}/f_T}$ is a cubic, its roots can certainly be obtained by extracting square roots and a cube root over $\mathbf{C}(W)$ (see Exercise 4.6.3), but in fact the Lagrange Lemma says that only the cube root is necessary (see Exercise 4.6.4).

Now the octahedral inversion procedure is

(4.6.2) ALGORITHM. *To invert any specific equation $f_O(z) = w$ for $w \in \mathbf{C}$,*

1. *Find a root w' of $m_{f_T/f_O}(T; w)$.*
2. *Find a root w'' of $m_{f_{D_2}/f_T}(T; w')$.*
3. *Find a root w''' of $m_{f_{C_2}/f_{D_2}}(T; w'')$.*
4. *Finally, find a root z_0 of $m_{Z/f_{C_2}}(T; w''')$. The z-values over w are then $\Gamma_O(z_0)$.*

Skipping the first step and taking $\Gamma_T(z_0)$ *in the last inverts the equation* $f_T(z) = w'$.

Calculating the rational relations among f_O, f_T, and f_{D_2} was undeniably tricky. Exercise 4.6.7 obtains the relations more naturally by geometric, function theoretic methods, and the next section will introduce some algebraic ideas to explain what's going on. The algebra will also make inroads in the icosahedral case.

Exercises

4.6.1. (a) Here is an algebraic proof that the alternating group A_n is **simple** (has no nontrivial normal subgroups) for $n \geq 5$. Fill in details as necessary. Suppose $\{1\} \neq K \triangleleft A_n$. As in Exercise 3.4.4, showing K contains a 3-cycle shows that $K = A_n$. The claim is that any nonidentity $g \in K$ that fixes a maximal number of $\{1, \ldots, n\}$ is in fact a 3-cycle. Otherwise g is one of the following disjoint cycle products: (a) $g = (1\,2\,3\,\cdots)\,\cdots$, or (b) $g = (1\,2)(3\,4)\,\cdots$, a product of transpositions. In case (a), $g \neq (1\,2\,3\,4)$, so g moves each of $\{1, 2, 3, 4, 5\}$. In either case, take $g_1 = (3\,4\,5)g(3\,5\,4)$ (right-to-left), obtaining $(1\,2\,4\,\cdots)\,\cdots$ in case (a) and $(1\,2)(4\,5)\,\cdots$ in case (b). In both cases $g_2 = g_1 g^{-1}$ is not the identity and fixes more elements than g, contradiction.

(b) Here is a geometric proof that the icosahedral group $\Gamma_I \cong A_5$ is simple. Let H be a normal subgroup. If H contains any element of order two, it contains all 15 such elements in Γ_I since all mid-edge rotations are conjugate. Similarly, if H contains any element of order three it contains all 20 such, and if H contains any element of order five then it contains all 24 such (this last case is a little trickier to argue than the first two—rotations by $2\pi/5$ and $4\pi/5$ are not conjugate). Thus

$$|H| = 1 + 15a + 20b + 24c \quad \text{with } a, b, c \in \{0, 1\}.$$

Since $|H|$ divides 60, each of a, b, c must be zero and H is trivial.

4.6.2. Derive the polynomials m_{f_T/f_O}, $m_{f_{D_2}/f_T}$, $m_{f_{C_n}/f_{D_n}}$, and $m_{Z/f_{C_n}}$ from the relevant rational relations.

4.6.3. Here is how to find the roots of the general cubic $f(T) = T^3 + bT^2 + cT + d \in \mathbf{k}(b, c, d)[T]$ where $\mathbf{k}$ is a field of characteristic other than 2 or 3, the symbols b, c, d can be numerical values or unknowns, and $\mathbf{k}(b, c, d)$ contains a primitive cube root of unity. (The procedure is given here without motivation. After we develop more theory the cubic will be solved again, systematically, in Exercise 5.1.6. The quartic is treated in Exercise 5.1.7.)

Step I: Substitute $T = X - b/3$ to reduce the problem to finding the roots of the **depressed** cubic polynomial $g(X) = X^3 + pX + q$ for certain $p, q \in \mathbf{k}(b, c, d)$.

Step II: Introduce new unknowns U, V such that $X = U + V$ and $UV = -p/3$. This gives the pair of conditions $U^3 + V^3 = -q$, $U^3V^3 = -p^3/27$. Therefore U^3 and V^3 are roots of the quadratic polynomial $Y^2 + qY - p^3/27$ and can be found by taking a square root.

Step III: Find possible pairs (U, V) by taking cube roots of U^3 and V^3 subject to the condition $UV = -p/3$. Substituting back $X = U + V$ and $T = X - b/3$ now gives the roots.

(a) Justify the steps as necessary. Where are the exceptional characteristics 2 and 3 disallowed?

(b) In step I, compute p and q in terms of b, c, d, perhaps with the help of a symbolic algebra package.

(c) In step II, assuming -3 has a square root in $\mathbf{k}(b, c, d)$, show that the square root operation is unnecessary if and only if $-4p^3 - 27q^2$ is a square in $\mathbf{k}(b, c, d)$. This quantity is called the **discriminant** of g.

(d) Find the roots over $\mathbf{C}$ of the polynomial $T^3 - 3T + 1$.

4.6.4. The Lagrange Lemma guarantees that finding the roots of $m_{f_{D_2}/f_T}$ doesn't require square roots. Use a symbolic algebra package to confirm that indeed the discriminant of $m_{f_{D_2}/f_T}$ is a square in $\mathbf{C}(W)$.

4.6.5. Find all $z \in \widehat{\mathbf{C}}$ such that $f_{D_n}(z) = 1/2$.

4.6.6. Find all $z \in \widehat{\mathbf{C}}$ such that $f_T(z) = \zeta_3^2$.

4.6.7. This exercise uses geometry and function theory to recalculate f_O as a rational expression in f_T, and f_T as a rational expression in f_{D_2}, without relying so heavily on formula crunching. Fill in the details as necessary.

For each noncyclic group Γ, the function f_Γ has been concocted to take the value ∞ at the vertex orbit $\mathcal{O}_1$ to order $n_1 = |\Gamma|/|\mathcal{O}_1|$; the value 0 at the face-center orbit $\mathcal{O}_2$ to order $n_2 = |\Gamma|/|\mathcal{O}_2|$; and the value 1 at the mid-edge orbit $\mathcal{O}_3$ to order $n_3 = |\Gamma|/|\mathcal{O}_3| = 2$. Each other value is taken by f_Γ on one full orbit to order 1.

In particular, the octahedral function f_O has poles of order 4 at the octahedral vertices and zeros of order 3 at the octahedral face-centers. To construct a rational expression in f_T that mimics this behavior, start with $1/(f_T - 1)^2$, which has poles of order 4 at the tetrahedral mid-edges—which are precisely the octahedral vertices—and zeros of order 6 at the tetrahedral vertices. Similarly, f_T has poles of order 3 at the tetrahedral vertices and zeros of order 3 at the tetrahedral face-centers. Thus the product $f_T/(f_T - 1)^2$ has the same zeros and poles as f_O, so $f_O = kf_T/(f_T - 1)^2$ for some $k \in \mathbf{C}$. Evaluating both sides of the equality at an octahedral mid-edge such as ζ_8 shows that $k = -4$.

As just mentioned, the tetrahedral function f_T has poles of order 3 at the tetrahedral vertices and zeros of order 3 at the tetrahedral face-centers. The tetrahedral face-centers form a D_2-orbit, so for any tetrahedral face-center c the difference $f_{D_2} - f_{D_2}(c)$ is independent of c and works out to $f_{D_2} + \zeta_3$. This has zeros of order 1 at the tetrahedral face-centers and poles of order 2 at the dihedral vertices $\{0, \infty\}$. Similarly for any tetrahedral vertex v, $f_{D_2} - f_{D_2}(v) = f_{D_2} + \zeta_3^2$. This has zeros of order 1 at the tetrahedral vertices and poles of order 2 at $\{0, \infty\}$. Thus $f_T = k(f_{D_2} + \zeta_3)^3/(f_{D_2} + \zeta_3^2)^3$ for some $k \in \mathbf{C}$. Evaluating both sides at the tetrahedral and dihedral mid-edge 1 shows $k = 1$.

7. Resolvents

The preceding section showed how to find z satisfying $f_\Gamma(z) = w$ for nonicosahedral Γ, via successive adjunctions of polynomial roots. Field-theoretically, the adjunctions constructed $\mathbf{C}(Z)$ from $\mathbf{C}(W)$. The polynomials $m(T; W)$ used to build intermediate fields between $\mathbf{C}(W)$ and $\mathbf{C}(Z)$ were obtained from certain rational relations, but they can also be obtained directly via a general procedure as follows.

Let $\mathbf{K}/\mathbf{k}$ be a Galois extension with group G. Take an element $r \in \mathbf{K}$ that is fixed by the subgroup $H \subset G$. Although H need not be normal in G, the right coset space $H\backslash G = \{H\gamma : \gamma \in G\}$ makes sense. It contains $[G : H]$ elements. Every element $h\gamma$ of the coset $H\gamma$ acts on r from the right as γ does, so for the purposes of acting on r, we may specify the coset space $H\backslash G$ by a set of representatives γ. The **resolvent** of r is the polynomial

$$R_{r,\mathbf{K}/\mathbf{k}} = \prod_{\gamma \in H\backslash G} (T - r^\gamma).$$

This polynomial is G-invariant, since as γ runs through a set of coset representatives so does γg for any fixed $g \in G$, giving $R^g_{r,\mathbf{K}/\mathbf{k}} = \prod_{\gamma \in H\backslash G}(T - r^{\gamma g}) =$

$R_{r,\mathbf{K}/\mathbf{k}}$. Therefore $R_{r,\mathbf{K}/\mathbf{k}}$ lies in $\mathbf{k}[T]$ by Galois theory. It is irreducible since G acts transitively on its roots, so it is the minimal polynomial of r over $\mathbf{k}$. The various extensions $\mathbf{k}(r^\gamma)/\mathbf{k}$ obtained by adjoining resolvent roots therefore have degree $[\mathbf{k}(r^\gamma):\mathbf{k}] = \deg(R_{r,\mathbf{K}/\mathbf{k}}) = [G:H]$. (See Exercise 4.7.1 for a description of the fixing subgroups of each $\mathbf{k}(r^\sigma)$ and of $\mathrm{spl}_{\mathbf{k}}(R_{r,\mathbf{K}/\mathbf{k}})$.) Thus on the one hand, the resolvent lets us write down the minimal polynomial of a specified r once we know its fixing subgroup H, and on the other hand, by judicious choice of r with appropriate fixing subgroup H we can construct specified intermediate fields between $\mathbf{k}$ and $\mathbf{K}$.

For a familiar example, recall the Galois extension $\mathbf{Q}(\zeta_5)/\mathbf{Q}$ with group $\langle\sigma:\zeta_5\mapsto\zeta_5^2\rangle$ cyclic of order 4. The element $\zeta_5+\zeta_5^{-1}\in\mathbf{Q}(\zeta_5)$ is fixed by the subgroup $\langle\sigma^2\rangle$ of the Galois group $\langle\sigma\rangle$. Representatives for $\langle\sigma^2\rangle\backslash\langle\sigma\rangle$ are $\{1,\sigma\}$, so

$$R_{\zeta_5+\zeta_5^{-1},\mathbf{Q}(\zeta_5)/\mathbf{Q}} = (T-(\zeta_5+\zeta_5^{-1}))(T-(\zeta_5^2+\zeta_5^{-2})).$$

Since $\zeta_5^5=1$ and $\sum_{i=0}^4\zeta_5^i=0$, this works out to T^2+T-1, again recovering that $\zeta_5+\zeta_5^{-1}$, being positive, is the golden ratio g.

For another example, reconsider the dihedral extension $\mathbf{C}(Z)/\mathbf{C}(f_{D_n})$. The cyclic invariant $f_{C_n}\in\mathbf{C}(Z)$ is fixed by the subgroup C_n of the Galois group D_n. Representatives for $C_n\backslash D_n$ are $\{1,t_D\}$ where $t_D(z)=1/z$, so

$$\begin{aligned}R_{f_{C_n},\mathbf{C}(Z)/\mathbf{C}(f_{D_n})} &= (T-f_{C_n})(T-f_{C_n}\circ t_D) = (T-Z^n)(T-1/Z^n)\\ &= T^2-(Z^n+1/Z^n)T+1.\end{aligned}$$

This reduces, as expected, to the familiar $m_{f_{C_n}/f_{D_n}}(T;W) = T^2+(2-4W)T+1$ where $W=f_{D_n}=(Z^n+1)^2/(4Z^n)$.

Similarly, return to the octahedral extension $\mathbf{C}(Z)/\mathbf{C}(f_O)$. The tetrahedral invariant $f_T\in\mathbf{C}(Z)$ is fixed by the subgroup Γ_T of the Galois group Γ_O. Representatives for $\Gamma_T\backslash\Gamma_O$ are $\{1,s_4\}$ where $s_4(z)=iz$. Note that $f_T\circ s_4 = 1/f_T$ since s_4 exchanges the tetrahedral vertices and the tetrahedral faces, where $f_T=\infty$ and $f_T=0$ respectively, and preserves the tetrahedral edges, where $f_T=1$. Thus,

$$R_{f_T,\mathbf{C}(Z)/\mathbf{C}(f_O)} = (T-f_T)(T-1/f_T) = T^2-(f_T+1/f_T)T+1.$$

Having degree $24=|\Gamma_O|$ and being Γ_O-invariant, f_T+1/f_T is a fractional linear transformation of f_O. Since it takes the value 2 at the octahedral vertices (which are the tetrahedral edges), the value ∞ at the octahedral faces (which are the tetrahedral vertices and faces), and the value -2 at the octahedral edge ζ_8, it must be $f_T+1/f_T=-4/f_O+2$. Now setting $W=f_O$

in the resolvent recovers the polynomial $m_{f_T/f_O}(T;W) = T^2+(4/W-2)T+1$.

Exercises

4.7.1. Let $\mathbf{K}/\mathbf{k}$ be a Galois extension with group G, and let $r \in \mathbf{K}$ be fixed by $H \subset G$. Show that $\mathbf{k}(r^\gamma) = \mathbf{K}^{(\gamma^{-1}H\gamma)}$ for all $\gamma \in G$. Show that $\mathrm{spl}_{\mathbf{k}}(R_{r,\mathbf{K}/\mathbf{k}}) = \mathbf{K}^N$ where $N = \cap_{\gamma \in H\backslash G}(\gamma^{-1}H\gamma)$ is the largest normal subgroup of G contained in H.

4.7.2. Consider the Galois extension $\mathbf{Q}(\zeta_5)/\mathbf{Q}(\zeta_5+\zeta_5^{-1})$ with group $\langle \tau : \zeta_5 \mapsto \zeta_5^{-1}\rangle$ cyclic of order 2. Compute the resolvent $R_{\zeta_5,\mathbf{Q}(\zeta_5)/\mathbf{Q}(\zeta_5+\zeta_5^{-1})}$. Give an expression for ζ_5 in radicals over $\mathbf{Q}$.

4.7.3. Recompute the polynomial $m_{f_{D_2}/f_T}$ as a resolvent. (This is harder than the examples in the section.)

8. The Brioschi resolvent

The icosahedral extension $\mathbf{C}(Z)/\mathbf{C}(f_I)$ has Galois group Γ_I containing five tetrahedral subgroups, each stabilizing a configuration of golden rectangles as described in Section 2.5. Single out the stabilizing group $\widetilde{\Gamma}_T$ of the configuration originally set in the coordinate planes but then rotated about the x_2-axis in Section 2.7. Thus $\widetilde{\Gamma}_T = \alpha\Gamma_T\alpha^{-1}$ where α is the rotation. (See Figure 4.8.1.) The field of invariant functions $\mathbf{C}(Z)^{\widetilde{\Gamma}_T}$ is generated by $\tilde{f}_T = f_T \circ \alpha^{-1}$. The corresponding tower of fields is as follows.

$$\begin{array}{rll} \mathbf{C}(Z) & & \\ 12 & \widetilde{\Gamma}_T & \\ \mathbf{C}(\tilde{f}_T) & & \Gamma_I \\ 5 & & \\ \mathbf{C}(f_I) & & \end{array}$$

Any $\widetilde{\Gamma}_T$-invariant rational function r has a resolvent $R_{r,\mathbf{C}(Z)/\mathbf{C}(f_I)}$ of degree $[\Gamma_I : \widetilde{\Gamma}_T] = 60/12 = 5$. Fairly general geometric, function theoretic techniques, akin to Exercise 4.6.7, provide a recipe for computing such R, but instead we will use a more convenient form-based method.

Let Γ'_I and $\widetilde{\Gamma}'_T$ denote the lifts of Γ_I and $\widetilde{\Gamma}_T$ to $\mathrm{SL}_2(\mathbf{C})$ and suppose the form $\tilde{F} \in \mathbf{C}[Z_1, Z_2]$ is $\widetilde{\Gamma}'_T$-invariant with trivial character. The polynomial

$$R_{\tilde{F}} = \prod_{\gamma' \in \widetilde{\Gamma}'_T \backslash \Gamma'_I} (T - \tilde{F} \circ \gamma')$$

is then well-defined and has degree $[\Gamma'_I : \widetilde{\Gamma}'_T] = 5$. Its coefficients are Γ'_I-invariant forms with trivial character and therefore lie in $\mathbf{C}[F_{1,I}, F_{2,I}, F_{3,I}]$.

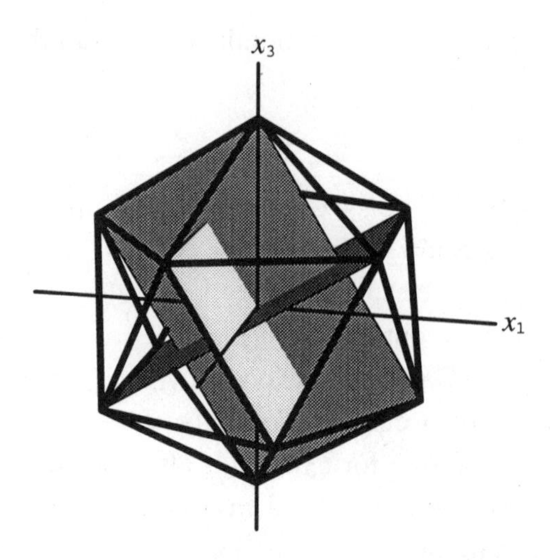

Figure 4.8.1. Rotated golden configuration

Certain rationalization techniques, better demonstrated explicitly than described in general, modify this **form resolvent** into a resolvent of the familiar type—that is, a polynomial in $\mathbf{C}(f_I)[T]$ with a $\widetilde{\Gamma}_T$-invariant rational function for one of its roots.

Obtaining $\widetilde{\Gamma}'_T$-invariant forms $\tilde{F}$, needed to compute form resolvents, from Γ'_T-invariant forms F, which we already have, is easy in principle. Lift the relation $\widetilde{\Gamma}_T = \alpha\Gamma_T\alpha^{-1}$ in $\mathrm{PSL}_2(\mathbf{C})$ to $\widetilde{\Gamma}'_T = \alpha'\Gamma'_T(\alpha')^{-1}$ in $\mathrm{SL}_2(\mathbf{C})$. If F is Γ'_T-invariant with character χ_F then Exercise 4.8.1 shows that the "rotated" form $\tilde{F} = F \circ (\alpha')^{-1}$ is $\widetilde{\Gamma}'_T$-invariant with essentially the same character, $\tilde{\chi}_{\tilde{F}}(\alpha\gamma\alpha^{-1}) = \chi_F(\gamma)$. (No primes are needed in the notation here because all Platonic characters descend to projective groups, cf. Section 3.4.) Since $\alpha^{-1} = f_{i,\arctan g}$ is described explicitly by Theorem 2.2.3, we can thus compute $\tilde{F}$, given F.

The simplest tetrahedral form with trivial character under $\widetilde{\Gamma}_T$ is $\tilde{F}_{3,T} = \tilde{F}_{1,O}$, the vertex form for the rotated octahedron (Figure 4.8.2). (In fact, since $\widetilde{\Gamma}_T = \langle \tilde{s}_O^2, \tilde{t}_O \rangle$, where in general, $\tilde{\gamma} = \alpha\gamma\alpha^{-1}$, inspection of the degenerate octahedral ground form characters shows that they are all trivial under $\widetilde{\Gamma}_T$. Thus each octahedral form provides a quintic icosahedral resolvent.)

Computing the resolvent for $\tilde{F}_{1,O}$ requires computing $\tilde{F}_{1,O} = F_{1,O} \circ (\alpha')^{-1}$ itself. A cute shortcut avoids computing $(\alpha')^{-1}$ and the composition. The idea is that the vertices of the rotated octahedron, where $\tilde{F}_{1,O}$ vanishes, come in three pairs: $\{\pm i\}$; the fixed points of the icosahedral rotation t_I; and the fixed points of $f_{i,\pi/2}^{-1} \circ t_I \circ f_{i,\pi/2}$, where the notation $f_{i,\pi/2}$ is as in

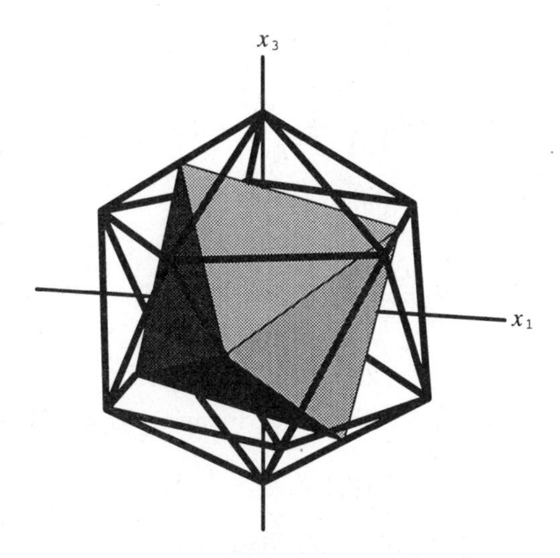

Figure 4.8.2. Rotated octahedron

Section 2.2. The form vanishing at $\{\pm i\}$ is $Z_1^2 + Z_2^2$. The fixed points of t_I satisfy $(g-z)/(gz+1) = z$, i.e., $z^2+(2/g)z-1=0$, so the form vanishing at such points is $Z_1^2+(2/g)Z_1Z_2-Z_2^2$. The fixed points of $f_{i,\pi/2}^{-1} \circ t_I \circ f_{i,\pi/2}$ satisfy $t_I(f_{i,\pi/2}(z)) = f_{i,\pi/2}(z)$. Theorem 2.2.3 reduces this to $t_I((z-1)/(z+1)) = (z-1)/(z+1)$ (Exercise 4.8.2), or equivalently, to $((z-1)/(z+1))^2 - (2/g)(z-1)/(z+1) - 1 = 0$, which works out to $z^2 - 2gz - 1 = 0$, so the corresponding form is $Z_1^2 - 2gZ_1Z_2 - Z_2^2$. Multiplying the three forms together gives (Exercise 4.8.3)

$$\tilde{F}_{1,O} = Z_1^6 + 2Z_1^5Z_2 - 5Z_1^4Z_2^2 - 5Z_1^2Z_2^4 - 2Z_1Z_2^5 + Z_2^6.$$

Note the slightly asymmetric distribution of coefficients.

Since $\deg(\tilde{F}_{1,O}) = 6$, its form resolvent

$$R_{\tilde{F}_{1,O}} = T^5 + a_1T^4 + a_2T^3 + a_3T^2 + a_4T + a_5$$

has as each coefficient a_i a form of total degree $6i$ in $\mathbf{C}[F_{1,I}, F_{2,I}, F_{3,I}]$. Since $\deg(F_{1,I}) = 12$, $\deg(F_{2,I}) = 20$ and $\deg(F_{3,I}) = 30$, it follows immediately that $a_1 = a_3 = 0$, $a_2 = k_2F_{1,I}$, $a_4 = k_4F_{1,I}^2$ and $a_5 = k_5F_{3,I}$. In the relation $R_{\tilde{F}_{1,O}}(\tilde{F}_{1,O}) = 0$, setting the coefficients of Z_1^{30}, $Z_1^{29}Z_2$ and $Z_1^{28}Z_2^2$ of the left side to zero gives

$$R_{\tilde{F}_{1,O}} = T^5 - 10F_{1,I}T^3 + 45F_{1,I}^2T - F_{3,I}.$$

To turn this into a resolvent whose coefficients depend on one parameter, note that $\tilde{F}_{1,O}F_{1,I}^2/F_{3,I}$, a quotient of same-degree rotated tetrahedral forms with trivial character, is a root of

$$\left(\frac{TF_{3,I}}{F_{1,I}^2}\right)^5 - 10F_{1,I}\left(\frac{TF_{3,I}}{F_{1,I}^2}\right)^3 + 45F_{1,I}^2\left(\frac{TF_{3,I}}{F_{1,I}^2}\right) - F_{3,I} = 0.$$

Equivalently, after some algebra,

$$T^5 - 10\left(\frac{F_{1,I}^5}{F_{3,I}^2}\right)T^3 + 45\left(\frac{F_{1,I}^5}{F_{3,I}^2}\right)^2 T - \left(\frac{F_{1,I}^5}{F_{3,I}^2}\right)^2 = 0.$$

Since $f_I = 1 - (F_{3,I}^2)_*/(1728F_{1,I}^5)_*$, setting $W' = 1/(1728(1-f_I))$ gives the resolvent of the rational tetrahedral invariant $\tilde{s} = (\tilde{F}_{1,O}F_{1,I}^2)_*/(F_{3,I})_*$,

$$R_{\tilde{s},\mathbf{C}(Z)/\mathbf{C}(f_I)}(T;W') = T^5 - 10W'T^3 + 45W'^2T - W'^2.$$

The right side of this last equality is the very important **Brioschi quintic**. Chapter 5 will discuss it and reduce solving the general quintic to the Brioschi case.

Exercise 4.8.5 mimics the techniques of this section to compute another quintic icosahedral resolvent, which also is in a form to be discussed in Chapter 5, and Exercise 4.8.6 sketches how to compute an entire family of such resolvents.

Exercises

4.8.1. Let $\tilde{\Gamma}'_T = \alpha'\Gamma'_T(\alpha')^{-1}$ as in the text. If F is a Γ'_T-invariant form with character χ_F, show that the rotated form $\tilde{F} = F \circ (\alpha')^{-1}$ is $\tilde{\Gamma}'_T$-invariant with essentially the same character, $\tilde{\chi}_{\tilde{F}}(\alpha\gamma\alpha^{-1}) = \chi_F(\gamma)$.

4.8.2. Show that $f_{i,\pi/2}(z) = (z-1)/(z+1)$ as claimed in the calculation of $\tilde{F}_{1,O}$.

4.8.3. Confirm the expression for $\tilde{F}_{1,O}$.

4.8.4. Confirm the formulas for the form resolvent $R_{\tilde{F}_{1,O}}$ and the resolvent $R_{\tilde{s}}$.

4.8.5. This exercise computes the resolvent for $\tilde{F}_{2,O}$. Take the Hessian of $\tilde{F}_{1,O}$ and divide by a factor to obtain

$$\tilde{F}_{2,O} = -Z_1^8 + Z_1^7Z_2^6 - 7Z_1^6Z_2^2 - 7Z_1^5Z_2^3 + 7Z_1^3Z_2^5 - 7Z_1^2Z_2^6 - Z_1Z_2^7 - Z_2^8.$$

Explain why the resolvent of this form is $R_{\tilde{F}_{2,O}} = T^5 + k_3F_{1,I}^2T^2 + k_4F_{1,I}F_{2,I}T + k_5F_{2,I}^2$ for some constants k_3, k_4, k_5. Equate coefficients in the relation $R_{\tilde{F}_{2,O}}(\tilde{F}_{2,O}) = 0$ to obtain

$$R_{\tilde{F}_{2,O}} = T^5 + 40F_{1,I}^2T^2 - 5F_{1,I}F_{2,I}T + F_{2,I}^2.$$

Homogenize $\tilde{F}_{2,O}$ to $\tilde{t} = (\tilde{F}_{2,O}F_{1,I})_*/(F_{2,I})_*$, set $W' = 1/(1728f_I)$, and show that $\tilde{t}$ is a root of

$$R_{\tilde{t},\mathbf{C}(Z)/\mathbf{C}(f_I)}(T;W') = T^5 + 40W'T^2 - 5W'T + W'.$$

This resolvent, with no T^4 or T^3 term, is in **principal form.**

4.8.6. In fact, the icosahedral equation has an entire family of principal quintic resolvents, one for each nonzero vector $(c_1, c_2) \in \mathbf{C}^2$. To see this, argue by degree that the resolvent

$$R_{c_1\tilde{F}_{2,O}+c_2\tilde{F}_{1,O}\tilde{F}_{2,O}} = \sum_{\gamma' \in \tilde{\Gamma}'_T \backslash \Gamma'_I} (T - (c_1\tilde{F}_{2,O} + c_2\tilde{F}_{1,O}\tilde{F}_{2,O}) \circ \gamma')$$

is principal. That is, its T^4 coefficient is a combination of Γ'_I-invariant forms of degrees 8 and 14, forcing it to be zero, and similarly for the T^3 coefficient.

The calculation may be skipped in the context of this book since we are more interested in the Brioschi quintic than in principal ones, but it works out to

$$\begin{aligned} R_{c_1\tilde{F}_{2,O}+c_2\tilde{F}_{1,O}\tilde{F}_{2,O}} &= T^5 + 5(8c_1^3F_{1,I}^2 + c_1^2c_2F_{3,I} \\ &\qquad + 72c_1c_2^2F_{1,I}^3 + c_2^3F_{1,I}F_{3,I})T^2 \\ &\quad + 5(-c_1^4F_{1,I}F_{2,I} + 18c_1^2c_2^2F_{1,I}^2F_{2,I} \\ &\qquad + c_1c_2^3F_{2,I}F_{3,I} + 27c_2^4F_{1,I}^3F_{2,I})T \\ &\quad + (c_1^5F_{2,I}^2 - 10c_1^3c_2^2F_{1,I}F_{2,I}^2 \\ &\qquad + 45c_1c_2^4F_{1,I}^2F_{2,I}^2 + c_2^5F_{2,I}^2F_{3,I}). \end{aligned}$$

Note that this degenerates to the previous problem when $(c_1, c_2) = (1, 0)$. As usual, this can be modified to a rational resolvent, whose coefficients will be functions of c_1, c_2 and f_I. Rationalize $c_1\tilde{F}_{2,O} + c_2\tilde{F}_{1,O}\tilde{F}_{2,O}$ to

$$\tilde{u} = d_1(12\tilde{F}_{2,O}F_{1,I})_*/(F_{2,I})_* + d_2(12\tilde{F}_{1,O}F_{1,I}^2)_*/(F_{3,I})_* \cdot (12\tilde{F}_{2,O}F_{1,I})_*/(F_{2,I})_*$$

where $d_1 = c_1(F_{2,I})_*/(12F_{1,I})_*$ and $d_2 = c_2(F_{2,I}F_{3,I})_*/(144F_{1,I}^3)_*$, and set $W' = 1/(1 - f_I)$; then $\tilde{u}$ satisfies the resolvent

$$\begin{aligned} R_{\tilde{u},\mathbf{C}(Z)/\mathbf{C}(f_I)}(T;W') &= T^5 + \frac{5W'}{W'-1}(8d_1^3 + 12d_1^2d_2 + (6d_1d_2^2 + d_2^3)W')T^2 \\ &\quad + \frac{15W'}{W'-1}(-4d_1^4 + (6d_1^2d_2^2 + 4d_1d_2^3)W' + \frac{3}{4}d_2^4W'^2)T \\ &\quad + \frac{3W'}{W'-1}(48d_1^5 - 40d_1^3d_2^2W' + (15d_1d_2^4 + 4d_2^5)W'^2). \end{aligned}$$

9. Inversion of the icosahedral invariant

Adjoining the root $\tilde{s} = (\tilde{F}_{1,O}F_{1,I}^2)_*/(F_{3,I})_*$ of the Brioschi resolvent to the ground field $\mathbf{C}(f_I)$ gives the field of rotated tetrahedral functions, $\mathbf{C}(\tilde{f}_T)$. (See Exercise 4.9.1.) Thus $\mathbf{C}(\tilde{f}_T) = \mathbf{C}(\tilde{s}, f_I)$, meaning that $\tilde{f}_T$ is a rational function of $\tilde{s}$ and f_I, or equivalently, a rational function of $\tilde{s}$ and the quantity $W' = 1/(1728(1 - f_I))$ from the preceding section. Finding this rational function will complete the icosahedral inversion since we know to invert the tetrahedral invariant.

A more conveniently expressed degree 12 tetrahedral invariant than $\tilde{f}_T$ is

$$\tilde{r} = \frac{(\tilde{F}_{1,O}^2)_*}{(F_{1,I})_*} = \frac{(\tilde{F}_{1,O}^2 F_{1,I}^4)_*}{(F_{3,I}^2)_*} \frac{(F_{3,I}^2)_*}{(F_{1,I}^5)_*} = \frac{\tilde{s}^2}{W'}$$

with explicit formula

$$\tilde{r} = \frac{(\tilde{F}_{1,O}^2)_*}{(F_{1,I})_*} = \frac{(Z^6 + 2Z^5 - 5Z^4 - 5Z^2 - 2Z + 1)^2}{Z(Z^{10} + 11Z^5 - 1)}.$$

The normalized invariant $\tilde{f}_T$, which we want, is a fractional linear transformation of $\tilde{r}$. Since $\tilde{r}$ vanishes at the edges of the rotated tetrahedron, the specific relation is

$$\tilde{f}_T = \frac{\tilde{r}(p)}{\tilde{r}(q)} \cdot \frac{\tilde{r} - \tilde{r}(q)}{\tilde{r} - \tilde{r}(p)}$$

where p is a tetrahedral vertex and q is a tetrahedral face. Exercise 4.9.2 shows that p and q may be taken as $\zeta_5^{-1}z$ and $\zeta_5 z$ respectively where $z = (2+g-\sqrt{9+3g})/2$. Exercise 4.9.4 gives $\tilde{r}(p) = (11+3i\sqrt{15})/2$, $\tilde{r}(q) = \overline{\tilde{r}(p)}$. Thus

$$\tilde{f}_T = \frac{11 + 3i\sqrt{15}}{11 - 3i\sqrt{15}} \cdot \frac{\tilde{r} - (11 - 3i\sqrt{15})/2}{\tilde{r} - (11 + 3i\sqrt{15})/2}. \tag{4.9.1}$$

Now the icosahedral inversion procedure is

(4.9.2) ALGORITHM. *To invert any specific icosahedral equation $f_I(z) = w$ with $w \in \mathbf{C}$,*

1. *Set $w' = 1/(1728(1 - w))$ and find a root $\tilde{s}$ of the Brioschi resolvent $R_{\tilde{s},\mathbf{C}(Z)/\mathbf{C}(f_I)}(T; w')$. Thus $\tilde{s} = \tilde{s}(z)$ for some value $z \in f_I^{-1}(w)$.*
2. *Set $\tilde{r} = \tilde{s}^2/w'$ and compute $w'' = \tilde{f}_T(z)$ from $\tilde{r}$ using (4.9.1).*
3. *Finally, since $\tilde{f}_T = f_T \circ \alpha^{-1}$, find a root z_0 of the equation $f_T(\alpha^{-1}z_0) = w''$ (cf. Section 4.6). The z-values over w are thus $\Gamma_I(z_0)$.*

Step 1, which requires a root $\tilde{s}$ of the Brioschi resolvent, is not in radicals.

Exercises

4.9.1. Explain how the tower of fields

$$\left.\begin{array}{c}\mathbf{C}(\tilde{f}_T)\\ \Big| \\ \mathbf{C}(\tilde{s}, f_I)\\ \Big| \\ \mathbf{C}(f_I)\end{array}\right\} 5$$

shows that either $\mathbf{C}(\tilde{s}, f_I) = \mathbf{C}(\tilde{f}_T)$ or $\mathbf{C}(\tilde{s}, f_I) = \mathbf{C}(f_I)$. Rule out the second possibility by considering the degree of $\tilde{s}$.

4.9.2. This exercise calculates the points p and q in the text. Explain why they may be taken as $p = \zeta_5^{-1} z$ and $q = \zeta_5 z$ where z is the icosahedral face-center of the leftmost inner triangle (with sides labeled 3, 4) in Figure 2.7.3. (It may help to find the point corresponding to z in Figure 4.9.1 and note that tetrahedral vertices sit at icosahedral face-centers.) Explain why $t_I(s_I^2 z) = s_I^{-2} z$. Compute from this condition, using the methods of Section 4.8, that $z = (2+g-\sqrt{9+3g})/2$. Where is $(2+g+\sqrt{9+3g})/2$ in Figure 4.9.1?

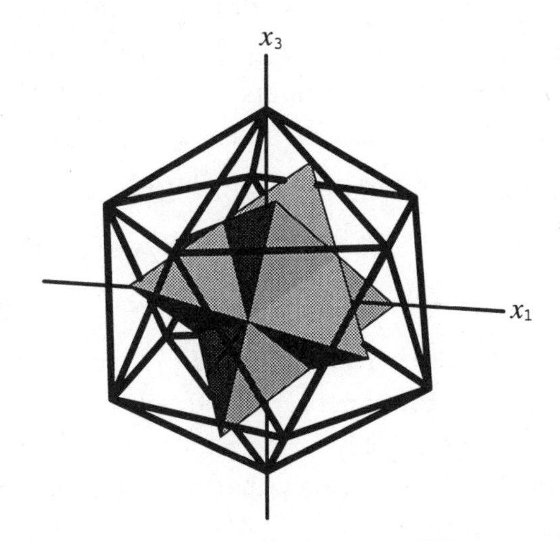

Figure 4.9.1. Tetrahedron and countertetrahedron

4.9.3. Explain why $\mathbf{C}(\tilde{s}) \neq \mathbf{C}(\tilde{r}) = \mathbf{C}(\tilde{f}_T)$. (See Theorem 3.6.2.) Divide the Brioschi relation $R_{\tilde{s},\mathbf{C}(Z)/\mathbf{C}(f_I)}(\tilde{s}) = 0$ by $W'^2 \tilde{s}$ to obtain

$$\tilde{r}^2 - 10\tilde{r} + 45 = 1/\tilde{s} \tag{4.9.3}$$

and conclude that $[\mathbf{C}(\tilde{r}) : \mathbf{C}(\tilde{s})] = 2$.

4.9.4. (a) Square both sides of (4.9.3) and multiply by $\tilde{r}$ to obtain a quintic equation satisfied by $\tilde{r}$ over $\mathbf{C}(W')$,

$$\tilde{r}(\tilde{r}^2 - 10\tilde{r} + 45)^2 = 1/W'.$$

Use the relation $1/W' = 1728(1 - f_I)$ and lots of high school algebra to show that consequently another quintic satisfied by $\tilde{r}$ is

$$(\tilde{r} - 3)^3(\tilde{r}^2 - 11\tilde{r} + 64) = -1728 f_I. \tag{4.9.4}$$

(b) Show that the 20 icosahedral face-centers decompose into three orbits under the rotated tetrahedral group $\widetilde{\Gamma}_T$: the four tetrahedral vertices in Figure 4.9.1 including p from Exercise 4.9.2, the four countertetrahedral vertices including q, and the remaining twelve icosahedral face-centers including z. (Hint: examine the stabilizer in $\widetilde{\Gamma}_T$ of each icosahedral face-center and use the Orbit-Stabilizer Theorem.) Since f_I vanishes at the icosahedral face-centers, (4.9.4) shows that the tetrahedral invariant $\tilde{r}$ takes the values $\left\{3, (11 \pm 3i\sqrt{15})/2\right\}$ there. Since the twelve-orbit contains the real-valued point z, $\tilde{r} = 3$ there. Verify the values $\tilde{r}(p)$ and $\tilde{r}(q)$ in the text by checking the sign of an imaginary part.

(c) Recall the rotated octahedral invariant $\tilde{F}_{2,O}$ from Exercise 4.8.5. Explain why the quotient $F_{2,I}/\tilde{F}_{2,O}$ lies in $\mathbf{C}[Z_1, Z_2]^{\widetilde{\Gamma}_T}$ and vanishes at the icosahedral face-centers where $\tilde{r} = 3$. (Compare the roots of the numerator and the denominator.) Compute that

$$\begin{aligned} F_{2,I}/\tilde{F}_{2,O} = Z_1^{12} &+ Z_1^{11}Z_2 - 6Z_1^{10}Z_2^2 - 20Z_1^9Z_2^3 + 15Z_1^8Z_2^4 \\ &- 24Z_1^7Z_2^5 + 11Z_1^6Z_2^6 + 24Z_1^5Z_2^7 + 15Z_1^4Z_2^8 \\ &+ 20Z_1^3Z_2^9 - 6Z_1^2Z_2^{10} - Z_1^{11}Z_2 - Z_2^{12}. \end{aligned}$$

Show that $\tilde{r} - 3 = (F_{2,I}/\tilde{F}_{2,O})_*/(F_{1,I})_*$ since both are degree 12 $\widetilde{\Gamma}_T$-invariants with the same zeros and poles and the same value at, e.g., $Z = 1$. We will use this formula for $\tilde{r} - 3$ in Chapter 7.

4.9.5. Use Theorem 2.2.3 to show that the rotation $\alpha = f_{\pi(0,-1,0),\arctan g}$ is

$$\alpha = \begin{bmatrix} \sqrt{(1 + 1/\sqrt{2-g})/2} & \sqrt{(1 - 1/\sqrt{2-g})/2} \\ -\sqrt{(1 - 1/\sqrt{2-g})/2} & \sqrt{(1 + 1/\sqrt{2-g})/2} \end{bmatrix}.$$

4.9.6. Conversely to the procedure at the end of the section, how does one find the roots of a specific Brioschi resolvent $R_{\tilde{s}}(T; w')$, given the icosahedral inverse?

4.9.7. Describe how to invert the icosahedral equation, given the root $\tilde{t}$ of the principal resolvent from Exercise 4.8.5. Conversely, how does one solve a specific principal resolvent $R_{\tilde{t}}(T; w')$, given the icosahedral inverse?

10. Summary

For each rotation group Γ, inverting the the equation $f_\Gamma(z) = w$ can be recast algebraically as the field-theoretic problem of constructing $\mathbf{C}(Z)$ from $\mathbf{C}(W)$. Galois theory and resolvents show how to carry this out by a succession of root adjunctions for nonicosahedral Γ. In the icosahedral case, the problem is not solvable by radicals alone, but also requires solving the Brioschi quintic. Conversely, the icosahedral inverse suffices to solve the Brioschi quintic, which can not be done by radicals.

CHAPTER 5

Reduction of the quintic to Brioschi form

Chapter 4 showed that solving the icosahedral equation is equivalent (modulo adjunction of radicals) to solving its Brioschi resolvent, the quintic

$$b = T^5 - 10W'T^3 + 45W'^2T - W'^2.$$

This chapter reduces solving the general quintic equation to this same b and therefore to the icosahedral equation. Thus, while radicals do not suffice to solve the general quintic, radicals and an icosahedral inverse do. What makes the reduction to Brioschi form so impressive is that the coefficients of b depend only on the single complex parameter W', where the general monic quintic has five independent parameters.

Recommended reading: The general material in this chapter can be found in standard algebra texts. The material more specific to the quintic follows Dickson [Di]. Chapter II.1 of Klein [Kl] discusses the 19th-century history of the quintic.

1. The general polynomial extension

This chapter studies polynomials over the field $\mathbf{C}$ of complex numbers. Doing so requires working with purely symbolic variables. Let n be a positive integer, and let $r_1, \ldots, r_n$ be **algebraically independent** symbols over $\mathbf{C}$, meaning that there is no nonzero polynomial $h \in \mathbf{C}[T_1, \ldots, T_n]$ such that $h(r_1, \ldots, r_n) = 0$. For instance, an algebraically independent one-element set over $\mathbf{C}$ is simply a transcendental variable. The monic polynomial p with roots $r_1, \ldots, r_n$ expands as

$$p = \prod_{i=1}^{n}(T - r_i) = \sum_{j \in \mathbf{Z}} (-1)^j \sigma_j T^{n-j} \in \mathbf{C}(\sigma_1, \ldots, \sigma_n)[T]$$

whose coefficients are (up to sign) the **elementary symmetric functions** of $r_1, \ldots, r_n$,

$$\sigma_j = \sigma_j(r_1, \ldots, r_n) = \begin{cases} \sum_{1 \le i_1 < \cdots < i_j \le n} \prod_{k=1}^{j} r_{i_k} & \text{for } j \ge 0 \\ 0 & \text{for } j < 0. \end{cases}$$

Note the special cases $\sigma_0 = 1$ and $\sigma_j = 0$ for $j > n$. (If the σ_j are new to you, work Exercise 5.1.2 before reading any farther.) The product form of p shows that the σ_j are invariant under all permutations of $r_1, \ldots, r_n$. The field extension $\mathbf{C}(r_1, \ldots, r_n)/\mathbf{C}(\sigma_1, \ldots, \sigma_n)$ is Galois, being a splitting field extension in characteristic zero. Let G denote the corresponding Galois group. Certainly G may be viewed as a subgroup of the symmetric group S_n since every automorphism of $\mathbf{C}(r_1, \ldots, r_n)$ over $\mathbf{C}(\sigma_1, \ldots, \sigma_n)$ must permute the roots of p. On the other hand, every permutation of $r_1, \ldots, r_n$ defines an automorphism of $\mathbf{C}(r_1, \ldots, r_n)$ fixing $\mathbf{C}(\sigma_1, \ldots, \sigma_n)$, so S_n is a subgroup of G. Therefore $G = S_n$.

By Exercise 5.1.3, $\sigma_1, \ldots, \sigma_n$ are algebraically independent over $\mathbf{C}$ given that $r_1, \ldots, r_n$ are. Thus each σ_j may be assigned a complex value without constraint, and the polynomial p, whose coefficients are (essentially) the σ_j, may be viewed as the general polynomial of degree n over $\mathbf{C}$. The field $\mathbf{C}(\sigma_1, \ldots, \sigma_n)$ is called the **coefficient field** of p, and $\mathbf{C}(r_1, \ldots, r_n)$ is the splitting field. The coefficients are expressed in terms of the roots; the goal is to invert this situation and express the roots in terms of the coefficients. In general, if $\mathbf{k}$ is a field, define the Galois group $\mathrm{Gal}_{\mathbf{k}}(p)$ of a separable polynomial $p \in \mathbf{k}[T]$ to be the Galois group of its splitting field extension $\mathrm{spl}_{\mathbf{k}}(p)/\mathbf{k}$. Then we have shown

(5.1.1) THEOREM. *The Galois group of the general polynomial p of degree n over $\mathbf{C}$ is the full symmetric group S_n. That is, $Gal_{\mathbf{C}(\sigma_1, \ldots, \sigma_n)}(p) = S_n$.*

The **discriminant** of $r_1, \ldots, r_n$ (also called the discriminant of p) is

$$\Delta = \Delta(r_1, \ldots, r_n) = \Delta(p) = \prod_{1 \le i < j \le n} (r_i - r_j)^2.$$

Being visibly invariant under S_n, the discriminant lies in the coefficient field of p. For example, if $n = 2$ then

$$\Delta = (r_1 - r_2)^2 = (r_1 + r_2)^2 - 4r_1 r_2 = \sigma_1^2 - 4\sigma_2.$$

Trying similarly to analyze the case $n = 3$ quickly shows that expressing Δ in terms of the σ_j is not easy. (Answer: $\sigma_1^2\sigma_2^2 - 4\sigma_2^3 - 4\sigma_1^3\sigma_3 - 27\sigma_3^2 + 18\sigma_1\sigma_2\sigma_3$.)

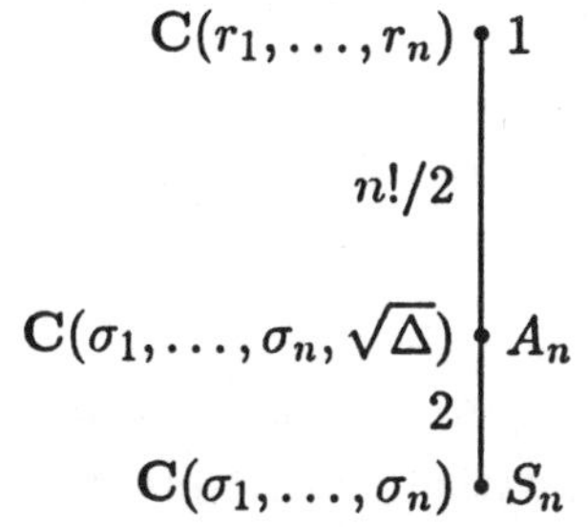

$$
\begin{array}{rcl}
\mathbf{C}(r_1,\ldots,r_n) & \bullet & 1 \\
n!/2 & | & \\
\mathbf{C}(\sigma_1,\ldots,\sigma_n,\sqrt{\Delta}) & \bullet & A_n \\
2 & | & \\
\mathbf{C}(\sigma_1,\ldots,\sigma_n) & \bullet & S_n
\end{array}
$$

Figure 5.1.1. The general polynomial extension

General expressions for the discriminant will be developed in the next two sections as byproducts of machinery.

Exercise 5.1.5 shows that the square root of the discriminant,

$$\sqrt{\Delta} = \prod_{1 \leq i < j \leq n} (r_i - r_j),$$

changes its sign when any two of the r's are exchanged, i.e., $\sqrt{\Delta}^{(k\,l)} = -\sqrt{\Delta}$ for any transposition $(k\,l) \in S_n$. So $\sqrt{\Delta}$ is fixed by A_n but not S_n, and the Galois Correspondence gives the tower shown in Figure 5.1.1. In field theoretic terms, the problem of solving the equation $p(T) = 0$ is tantamount to constructing the root field of p from the coefficient field. Adjoining the square root of the discriminant is the obvious first step, reducing the problem from an S_n extension to an A_n extension. Since A_n is solvable for $n \leq 4$, the techniques of the previous chapter will give solutions by radicals of the general cubic and quartic; see Exercises 5.1.6 and 5.1.7 for sketches of resolvent-based methods. For degree $n \geq 5$ the situation changes: the group A_n is simple (this was Exercise 4.6.1(a)), so by the Radical Criterion (Section 4.5) the general polynomial of degree n can not be solved by radicals.

Following some more generalities, this chapter discusses the case $n = 5$. After adjunction of the square root of the discriminant, the splitting field extension of the general quintic is an A_5 extension over $\mathbf{C}(\sigma_1, \ldots, \sigma_n, \sqrt{\Delta})$ with five parameters—$\sigma_1, \ldots, \sigma_5$ are algebraically independent, but along with $\sqrt{\Delta}$ they form an algebraically dependent set. On the other hand, inverting the icosahedral equation (or, equivalently, solving the Brioschi quintic) constructs $\mathbf{C}(Z)/\mathbf{C}(W)$, an A_5 extension over a field with only one parameter W. Reducing the general quintic to the Brioschi case will not decrease the Galois group of the extension, but it will diminish a situation depending on five parameters to a situation depending on only one, giving a considerable simplification.

Exercises

5.1.1. Let $r_1, \ldots, r_n$ be algebraically independent over $\mathbf{C}$. Prove that there is no nonzero rational function $f \in \mathbf{C}(T_1, \ldots, T_n)$ such that $f(r_1, \ldots, r_n) = 0$.

5.1.2. Write the elementary symmetric functions when $n = 1, 2, 3, 4$.

5.1.3. Prove that if $r_1, \ldots, r_n$ are algebraically independent over $\mathbf{C}$ then so are their elementary symmetric functions $\sigma_1, \ldots, \sigma_n$.

5.1.4. Let r_1, r_2, r_3 be algebraically related by $r_1 + r_2 + r_3 = 0$. Express the elementary symmetric functions $\sigma_1, \sigma_2, \sigma_3$ in terms of r_1 and r_2. Express the discriminant Δ in terms of $\sigma_1, \sigma_2, \sigma_3$.

5.1.5. Prove that the discriminant Δ is invariant under permutations of $r_1, \ldots, r_n$, but its square root satisfies $\sqrt{\Delta}^{(k\,l)} = -\sqrt{\Delta}$ for any transposition $(k\,l)$ of $r_1, \ldots, r_n$.

5.1.6. The **Fundamental Theorem of Symmetric Functions** says that any polynomial $p \in \mathbf{C}[r_1, \ldots, r_n]$ that is invariant under all permutations of $\{r_1, \ldots, r_n\}$ is expressible as a polynomial $q \in \mathbf{C}[\sigma_1, \ldots, \sigma_n]$. The proof is constructive (see, for example, [Co-Li-O'S]) and in particular when $n = 3$ yields the identities

$$(r_1 + \zeta_3 r_2 + \zeta_3^2 r_3)(r_1 + \zeta_3 r_3 + \zeta_3^2 r_2) = \sigma_1^2 - 3\sigma_2$$
$$(r_1 + \zeta_3 r_2 + \zeta_3^2 r_3)^3 + (r_1 + \zeta_3 r_3 + \zeta_3^2 r_2)^3 = 2\sigma_1^3 - 9\sigma_1\sigma_2 + 27\sigma_3.$$

Now we can solve the general cubic. Let $r = (r_1 + \zeta_3 r_2 + \zeta_3^2 r_3)$. Show that r^3 is invariant under A_3 but not S_3 so that its resolvent over the coefficient field $\mathbf{C}(\sigma_1, \sigma_2, \sigma_3)$ is the quadratic

$$\begin{aligned} R_{r^3} &= (T - r^3)(T - (r^3)^{(2\,3)}) \\ &= T^2 - (2\sigma_1^3 - 9\sigma_1\sigma_2 + 27\sigma_3)T + (\sigma_1^2 - 3\sigma_2)^3. \end{aligned}$$

Thus, taking a square root over the coefficient field gives r^3 and $(r^3)^{(2\,3)}$. (We don't know which is which because there is no canonical labeling of r_1, r_2, r_3, so just designate one as r^3.) Now r is a root of

$$R_r = T^3 - r^3$$

(there are three roots, but again they are indistinguishable under relabelling of the r_i), and $r^{(2\,3)} = (\sigma_1^2 - 3\sigma_2)/r$ from the first identity above. Now that we have r and $r^{(2\,3)}$, find r_1, r_2, and r_3 by solving the linear system

$$\begin{aligned} r_1 + \zeta_3 r_2 + \zeta_3^2 r_3 &= r \\ r_1 + \zeta_3^2 r_2 + \zeta_3 r_3 &= r^{(2\,3)} \\ r_1 + \quad r_2 + \quad r_3 &= \sigma_1. \end{aligned}$$

Use these methods to solve the cubic polynomial $T^3 - 3T + 1$ from Exercise 4.6.3.

5.1.7. Let $n = 4$, let $r = r_1 - r_2 + r_3 - r_4$, and let $s = r^2$. Show that the subgroup of S_4 leaving s invariant is the dihedral group $D = \langle (1\,2\,3\,4), (1\,3) \rangle$, and that a set of coset representatives for $D \backslash S_4$ is $\{1, (1\,2), (1\,4)\}$. The Fundamental Theorem of Symmetric Functions gives the identities

$$\begin{aligned} r \cdot r^{(1\,2)} \cdot r^{(1\,4)} &= \sigma_1^3 - 4\sigma_1\sigma_2 + 8\sigma_3 \\ s + s^{(1\,2)} + s^{(1\,4)} &= 3\sigma_1^2 - 8\sigma_2 \\ s \cdot s^{(1\,2)} + s \cdot s^{(1\,4)} + s^{(1\,2)} s^{(1\,4)} &= 3\sigma_1^4 - 16\sigma_1^2\sigma_2 + 16\sigma_1\sigma_3 + 16\sigma_2^2 - 64\sigma_4. \end{aligned}$$

To solve the quartic, take the cubic resolvent of s,

$$\begin{aligned} R_s &= (T - s)(T - s^{(1\,2)})(T - s^{(1\,4)}) \\ &= T^3 - (3\sigma_1^2 - 8\sigma_2)T^2 + (3\sigma_1^4 - 16\sigma_1^2\sigma_2 + 16\sigma_1\sigma_3 + 16\sigma_2^2 - 64\sigma_4)T \\ &\quad - (\sigma_1^3 - 4\sigma_1\sigma_2 + 8\sigma_3)^2. \end{aligned}$$

The three roots are s, $s^{(1\,2)}$, and $s^{(1\,4)}$; taking square roots of the first two gives r and $r^{(1\,2)}$, so by the first displayed identity, $r^{(1\,4)} = (\sigma_1^3 - 4\sigma_1\sigma_2 + 8\sigma_3)/(r \cdot r^{(1,2)})$. Now to solve the original quartic, solve the linear system

$$\begin{aligned} r_1 - r_2 + r_3 - r_4 &= r \\ -r_1 + r_2 + r_3 - r_4 &= r^{(1\,2)} \\ -r_1 - r_2 + r_3 + r_4 &= r^{(1\,4)} \\ r_1 + r_2 + r_3 + r_4 &= \sigma_1. \end{aligned}$$

2. Newton's identities

Retaining the notation from the the preceding section, define the **power sums** of $r_1, \ldots, r_n$ to be

$$s_j = s_j(r_1, \ldots, r_n) = \begin{cases} \sum_{i=1}^n r_i^j & \text{for } j \geq 0 \\ 0 & \text{for } j < 0 \end{cases}$$

including $s_0 = n$. These are clearly invariant under all permutations of r_1, ..., r_n. We want to relate them to the elementary symmetric functions σ_j. Start from the general polynomial,

$$p = \prod_{i=1}^{n}(T - r_i) = \sum_{j \in \mathbf{Z}}(-1)^j \sigma_j T^{n-j}.$$

Certainly $p' = \sum_{j\in\mathbf{Z}}(-1)^j\sigma_j(n-j)T^{n-j-1}$. But also, the logarithmic derivative and geometric series formulas,

$$\frac{p'}{p} = \sum_{i=1}^{n}\frac{1}{T - r_i} \qquad \text{and} \qquad \frac{1}{T-r} = \sum_{k=0}^{\infty}\frac{r^k}{T^{k+1}},$$

give

$$p' = p \cdot \frac{p'}{p} = p\sum_{i=1}^{n}\sum_{k=0}^{\infty}\frac{r_i^k}{T^{k+1}} = p\sum_{k\in\mathbf{Z}}\frac{s_k}{T^{k+1}} = \sum_{k,l\in\mathbf{Z}}(-1)^l\sigma_l s_k T^{n-k-l-1}$$

$$= \sum_{j\in\mathbf{Z}}\left[\sum_{l\in\mathbf{Z}}(-1)^l\sigma_l s_{j-l}\right]T^{n-j-1} \quad (\text{letting } j = k+l).$$

Equating the coefficients of the two expressions for p' gives the formula $\sum_{l=0}^{j-1}(-1)^l\sigma_l s_{j-l}+(-1)^j\sigma_j n = (-1)^j\sigma_j(n-j)$. **Newton's identities** follow,

$$\sum_{l=0}^{j-1}(-1)^l\sigma_l s_{j-l} + (-1)^j\sigma_j j = 0 \qquad \text{for all } j.$$

Explicitly, Newton's identities are

$$s_1 - \sigma_1 = 0$$

$$s_2 - s_1\sigma_1 + 2\sigma_2 = 0$$

$$s_3 - s_2\sigma_1 + s_1\sigma_2 - 3\sigma_3 = 0$$

$$s_4 - s_3\sigma_1 + s_2\sigma_2 - s_1\sigma_3 + 4\sigma_4 = 0$$

and so on.

These show (Exercise 5.2.4) that for any $j \in \{1, \ldots, n\}$, the power sums s_1, ..., s_j are polynomials (with no constant terms) in the elementary symmetric functions σ_1, ..., σ_j, and—since we are in characteristic zero—that the elementary symmetric functions σ_1, ..., σ_j are polynomials (with no constant terms) in the power sums s_1, ..., s_j. Consequently,

(5.2.1) PROPOSITION. *The first j coefficients a_1, ..., a_j of the polynomial $p = T^n + a_1T^{n-1} + \cdots + a_n$ are zero exactly when the first j power sums of its roots vanish.*

Exercises

5.2.1. Express s_j in terms of $\sigma_1, \ldots, \sigma_j$ for $j = 1, 2, 3$, and conversely.

5.2.2. Write some of Newton's identities when $j > n$; what is the pattern?

5.2.3. True or false: the second coefficient a_2 of the polynomial $p = T^n + a_1 T^{n-1} + \cdots + a_n$ is zero exactly when the second power sum of its roots vanishes.

5.2.4. Show that for any $j \in \{1, \ldots, n\}$, the power sums $s_1, \ldots, s_j$ are polynomials (with no constant terms) in the elementary symmetric functions $\sigma_1, \ldots, \sigma_j$, and conversely. (The converse fails in nonzero characteristic, for example consider $p = T^2 + 1$ in characteristic 2.)

5.2.5. Establish the formula for the **Vandermonde determinant**,

$$\begin{vmatrix} 1 & r_1 & r_1^2 & \cdots & r_1^{n-1} \\ 1 & r_2 & r_2^2 & \cdots & r_2^{n-1} \\ \vdots & \vdots & \vdots & & \vdots \\ 1 & r_n & r_n^2 & \cdots & r_n^{n-1} \end{vmatrix} = \prod_{i<j} (r_j - r_i).$$

(Replace the last column by $(p(r_1), \ldots, p(r_n))$ where $p = \prod_{i=1}^{n-1}(T - r_i)$.) Left-multiply the Vandermonde matrix by its transpose and take determinants to obtain

$$\begin{vmatrix} s_0 & s_1 & \cdots & s_{n-1} \\ s_1 & s_2 & \cdots & s_n \\ \vdots & \vdots & & \vdots \\ s_{n-1} & s_n & \cdots & s_{2n-2} \end{vmatrix} = \Delta(r_1, \ldots, r_n).$$

This expresses the discriminant in terms of the elementary symmetric functions $\sigma_1, \ldots, \sigma_n$ since Newton's identities give expressions for the power sums s_j in terms of the σ_j. A formula for Δ that doesn't require Newton's identities will be developed in the next section.

3. Resultants

Given polynomials p and q, we can determine whether they have a root in common without actually finding their roots. Let m and n be nonnegative integers, let $a_0, \ldots, a_m, b_0, \ldots, b_n$ be symbols (possibly elements of the base field $\mathbf{C}$) with $a_0 \neq 0$ and $b_0 \neq 0$, and let $\mathbf{k} = \mathbf{C}(a_0, \ldots, a_m, b_0, \ldots, b_n)$. The polynomials $p = \sum_{i=0}^{m} a_i T^{m-i}$ and $q = \sum_{i=0}^{n} b_i T^{n-i}$ in $\mathbf{k}[T]$ are utterly general when the a_i's and the b_i's form an algebraically independent set, or conversely they can be explicit polynomials when all the coefficients are in $\mathbf{C}$. By Exercise 5.3.1, the polynomials p and q share a nonconstant factor in $\mathbf{k}[T]$

if and only if there exist nonzero polynomials $P = \sum_{i=0}^{n-1} c_i T^{n-1-i} \in \mathbf{k}[T]$ of degree less than n and $Q = \sum_{i=0}^{m-1} d_i T^{m-1-i} \in \mathbf{k}[T]$ of degree less than m such that $pP = qQ$. Such P and Q exist if and only if the system

$$vM = 0$$

of $m + n$ linear equations over $\mathbf{k}$ in $m + n$ unknowns has a nonzero solution v, where $v = [c_0, c_1, \ldots, c_{n-1}, -d_0, -d_1, \ldots, -d_{m-1}]$ lies in $\mathbf{k}^{m+n}$, and M is the **Sylvester matrix**

$$M = \begin{bmatrix} a_0 & a_1 & \cdots & \cdots & a_m & & \\ & \ddots & \ddots & & & \ddots & \\ & & a_0 & a_1 & \cdots & \cdots & a_m \\ b_0 & b_1 & \cdots & b_n & & & \\ & b_0 & b_1 & \cdots & b_n & & \\ & & \ddots & \ddots & & \ddots & \\ & & & b_0 & b_1 & \cdots & b_n \end{bmatrix}$$

(n staggered rows of a_i's, m staggered rows of b_j's, all other entries 0), which lies in $\mathrm{M}_{m+n,m+n}(\mathbf{k})$. Such a nonzero solution exists in turn if and only if $\det M = 0$. This determinant, an element of $\mathbf{C}[a_0, \ldots, a_m, b_0, \ldots, b_n]$, is called the **resultant** of p and q and written $R(p, q)$. The condition that p and q share a factor in $\mathbf{k}[T]$ is equivalent to their sharing a root in the splitting field over $\mathbf{k}$ of pq. Thus the result is

(5.3.1) THEOREM. *The polynomials p and q in $\mathbf{k}[T]$ share a nonconstant factor in $\mathbf{k}[T]$, or equivalently, share a root in the splitting field over $\mathbf{k}$ of their product, if and only if $R(p, q) = 0$.*

When the coefficients of p and q are algebraically independent, $R(p, q)$ is a master formula that applies to all polynomials of degrees m and n. At the other extreme, when the coefficients are specific values in $\mathbf{C}$, $R(p, q)$ is a complex number that is zero or nonzero depending on whether the particular polynomials p and q share a factor.

Taking the resultant of p and q to check whether they share a root may also be viewed as eliminating the variable T from the pair of equations $p(T) = 0$ and $q(T) = 0$, leaving one equation $R(p, q) = 0$ in the remaining variables $a_0, \ldots, a_m, b_0, \ldots, b_n$.

In principle, evaluating $R(p, q) = \det M$ may be carried out via row and column operations. In practice, evaluating a large determinant is an error-prone process by hand. The next theorem will supply as a corollary a more efficient method to compute $R(p,q)$. In any case, since any worthwhile computer symbolic algebra package is equipped with a resultant function, nontrivial resultants can often be found by machine.

In their splitting field over $\mathbf{k}$, the polynomials p and q factor as

$$p = a_0 \prod_{i=1}^{m} (T - r_i), \qquad q = b_0 \prod_{j=1}^{n} (T - s_j).$$

To express the resultant $R(p,q)$ explicitly in terms of the roots of p and q introduce the quantity $\tilde{R}(p,q) = a_0^n b_0^m \prod_{i=1}^m \prod_{j=1}^n (r_i - s_j)$. This polynomial vanishes if and only if p and q share a root, so it divides $R(p,q)$. Note that $\tilde{R}(p,q)$ is homogeneous of degree mn in the r_i and s_j. On the other hand, each coefficient $a_i = a_0(-1)^i \sigma_i(r_1, \ldots, r_m)$ of p has homogeneous degree i in $r_1, \ldots, r_m$, and similarly for each b_j and $s_1, \ldots, s_n$. Thus in the Sylvester matrix the (i,j)th entry has degree

$$\begin{cases} j - i \text{ in the } r_i & \text{if } 1 \le i \le n,\ i \le j \le i+m, \\ j - i + n \text{ in the } s_j & \text{if } n+1 \le i \le n+m,\ i-n \le j \le i. \end{cases}$$

It quickly follows that any nonzero term in the determinant $R(p,q)$ has degree mn in the r_i and the s_j, so $\tilde{R}(p,q)$ and $R(p,q)$ agree up to multiplicative constant. Matching coefficients of $(s_1 \cdots s_n)^m$ shows that the constant is 1. This proves

(5.3.2) THEOREM. *The resultant of the polynomials* $p = \sum_{i=0}^m a_i T^{m-i} = a_0 \prod_{i=1}^m (T - r_i)$ *and* $q = \sum_{j=0}^n b_j T^{n-j} = b_0 \prod_{j=1}^n (T - s_j)$ *is given by the formulas*

$$R(p,q) = a_0^n b_0^m \prod_{i=1}^{m} \prod_{j=1}^{n} (r_i - s_j) = a_0^n \prod_{i=1}^{m} q(r_i) = (-1)^{mn} b_0^m \prod_{j=1}^{n} p(s_j).$$

A special case of this theorem gives the efficient formula for the discriminant promised earlier. See Exercise 5.3.4.

Computing resultants can now be carried out via a Euclidean algorithm procedure: repeatedly do polynomial division with remainder and apply formula (4) in

(5.3.3) COROLLARY. *The following formulas hold:*

1. $R(q,p) = (-1)^{mn} R(p,q)$.
2. $R(p\tilde{p}, q) = R(p,q) R(\tilde{p}, q)$ *and* $R(p, q\tilde{q}) = R(p,q) R(p, \tilde{q})$.
3. $R(a_0, q) = a_0^n$ *and* $R(a_0 T + a_1, q) = a_0^n q(-a_1/a_0)$.

4. If $q = Qp + \tilde{q}$ *with* $\deg(\tilde{q}) < \deg(p)$ *then*

$$R(p,q) = a_0^{\deg(q)-\deg(\tilde{q})} R(p,\tilde{q}).$$

Exercise 5.3.5 asks for the proofs.

Exercises

5.3.1. Show that p and q share a nonconstant factor in $\mathbf{k}[T]$ if and only if there exist nonzero polynomials P of degree less than n and Q of degree less than m in $\mathbf{k}[T]$ such that $pP = qQ$.

5.3.2. Write out the matrix M for various small values of m and n, and compute the corresponding resultants.

5.3.3. Fill in the details of the proof of Theorem 5.3.2.

5.3.4. (a) Use Theorem 5.3.2 to show that if p is monic, so that $p' = \sum_{i=1}^{n} \prod_{j \neq i} (T - r_j)$, then $R(p,p') = (-1)^{n(n-1)/2}\Delta(p)$.
(b) Use part (a) to recompute the discriminants of $p = T^2 + bT + c$ and of $p = T^3 + bT + c$.

5.3.5. (a) Prove the formulas in Corollary 5.3.3.
(b) Let $p = T^n + bT + c$. Compute $\Delta(p) = (-1)^{n(n-1)/2} R(p,p')$ using the corollary. (Do a polynomial division and apply the second formula in (3). Answer: $(-1)^{(n-1)(n-2)/2}(n-1)^{n-1}b^n + (-1)^{n(n-1)/2}n^n c^{n-1}$.) Note that since n is a general symbol here, evaluating $R(p,p')$ as a determinant is much more awkward than this method.

4. Tschirnhaus transformations and principal form

Let $a_1, \ldots, a_n$ be symbols (possibly elements of the base field $\mathbf{C}$), and let $\mathbf{k} = \mathbf{C}(a_1, \ldots, a_n)$. A **Tschirnhaus transformation** is a substitution that reduces finding the roots of the polynomial

$$p = T^n + a_1 T^{n-1} + \cdots + a_n = \prod_{i=1}^{n} (T - r_i) \in \mathbf{k}[T]$$

to finding the roots of another—possibly more tractable—polynomial q, and solving an auxiliary polynomial equation. Specifically, a Tschirnhaus transformation is a polynomial $t \in \mathbf{l}[T]$ with coefficients in a yet-unspecified extension field $\mathbf{l}$ of $\mathbf{k}$, possibly $\mathbf{l} = \mathbf{k}$. (The nature of $\mathbf{l}$ will be clarified via examples in this section and discussed further in Section 6.1.) Given p and t, define a new polynomial q by transforming the roots of p via t,

$$q = \prod_{i=1}^{n} (S - t(r_i)) = S^n + c_1 S^{n-1} + \cdots + c_n.$$

This monic polynomial has degree n in S, and since its coefficients are $c_j = (-1)^j \sigma_j(\{t(r_i)\})$, it lies in $\mathbf{l}[S]$, i.e., it is defined over the Tschirnhaus coefficient field. The condition $q(S) = 0$ is equivalent to existence of a T such that $S = t(T)$ and $p(T) = 0$, so q is explicitly calculable up to constant multiple as the resultant obtained by eliminating T,

$$q = cR(p, t - S) \in \mathbf{l}[S] \qquad \text{for some } c \in \mathbf{l}.$$

(Exercise 5.4.1(a) asks you to confirm this, and Exercise 5.4.1(b) gives another method for computing q.) Finding a root r of p now reduces to a two–step process:

1. Find a root s of q.
2. Find all values of r satisfying the auxiliary equation $t(r) = s$ and substitute them into p; then $p(r) = 0$ for at least one such r.

The idea of the Tschirnhaus transformation is to choose the substitution t to move the roots so that each of these steps is more convenient than solving p. The term "convenient" is context-dependent; it is further discussed under one circumstance in Section 6.1. In any case, the transformed polynomial q only depends on the values $t(r_i)$, so the division algorithm

$$t(r_i) = Q(r_i)p(r_i) + R(r_i) = R(r_i) \qquad \text{for some } Q, R \in \mathbf{l}[T],\ \deg(R) < \deg(p)$$

shows that without loss of generality $\deg(t) < \deg(p)$.

Two useful elementary Tschirnhaus transformations will illustrate matters. The easiest transformation is an affine map that reduces the general polynomial of degree n to **depressed** form, meaning the coefficient of T^{n-1} is zero. The procedure is clear: since the coefficient of T^{n-1} is (minus) the sum of the roots, translating the roots by their average gives new roots that sum to zero. In Tschirnhaus terms, we are given

$$p = T^n + a_1 T^{n-1} + \cdots + a_n = \prod_{i=1}^{n} (T - r_i)$$

and want to choose the transformation $t_1 = T + b_1$ so that the resulting

$$q = \prod_{i=1}^{n} (S - t_1(r_i)) = S^n + c_1 S^{n-1} + \cdots + c_n$$

has first coefficient $c_1 = 0$. A computation (Exercise 5.4.2) shows that $c_1 = -\sigma_1 - nb_1$ where $\sigma_1 = \sigma_1(r_1, \ldots, r_n)$ is the first symmetric function of the roots of p, so that the appropriate substitution is indeed $t_1 = T + a_1/n$.

Note that this example reduces solving the quadratic equation to taking a square root.

Exercise 5.4.3 similarly finds a quadratic Tschirnhaus transformation $t_2 = T^2 + b_1 T + b_2$ that reduces the depressed polynomial p of degree n to **principal** form, meaning the coefficients of T^{n-1} and T^{n-2} are zero. Computation shows that this requires the Tschirnhaus coefficient b_1 to satisfy the quadratic condition

$$s_2 b_1^2 + 2 s_3 b_1 + s_4 - s_2^2/n = 0,$$

where the s_i are the power sums of the roots of p. Thus, constructing b_1 requires a square root over the coefficient field of p. This typifies why Tschirnhaus coefficients are allowed to lie in an extension field $\mathbf{l}$ of $\mathbf{k}$ rather than being constrained to $\mathbf{k}$. The Tschirnhaus coefficient b_2 works out to $-s_2/n$. In particular, since a principal cubic is solved by taking a cube root, the transformations t_1 and t_2 combine to solve the cubic equation.

Field-theoretically, a Tschirnhaus transformation t gives rise to the following situation: as before, let $\mathbf{k}$ be the coefficient field of a polynomial p. Let $\mathbf{K} = \mathrm{spl}_{\mathbf{k}}(p) = \mathbf{k}(r_1, \ldots, r_n)$ where the r_i are the roots of p. Since we are working over the field $\mathbf{C}$ of characteristic zero, $\mathbf{K}/\mathbf{k}$ is Galois; let $G = \mathrm{Gal}(\mathbf{K}/\mathbf{k})$. Let $\mathbf{l}$ be the extension of $\mathbf{k}$ generated by the coefficients of t, let $\mathbf{L} = \mathbf{l}(r_1, \ldots, r_n) = \mathbf{Kl}$ (the field generated by $\mathbf{K}$ and $\mathbf{l}$), and let $\Gamma = \mathrm{Gal}(\mathbf{L}/\mathbf{l})$. Figure 5.4.1 depicts the situation.

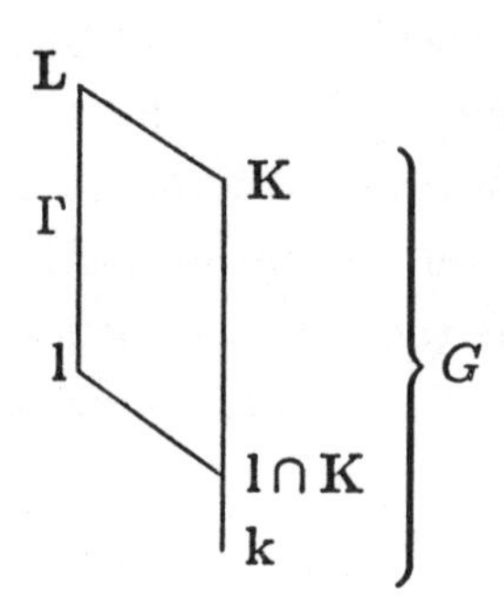

Figure 5.4.1.

This raises some obvious questions:

Is the new coefficient field $\mathbf{l}$ a proper extension of $\mathbf{k}$? For the affine transformation t_1 above, $\mathbf{l}$ is just $\mathbf{k}$. On the other hand, the coefficients of the quadratic transformation t_2 above require a square root that could make $\mathbf{l}$ a degree 2 extension of $\mathbf{k}$.

When $\mathbf{l}$ is a proper extension of $\mathbf{k}$, is it external to $\mathbf{K}$, the splitting field we are trying to construct? Exercise 5.4.4 shows that in the case of t_2, when $n = 3$ (i.e., the original polynomial p is the general cubic), the new coefficient field $\mathbf{l}$ is generated over $\mathbf{k}$ by the square root of the discriminant of p, so in fact $\mathbf{l} \subset \mathbf{K}$. But for general n this is not the case: the coefficient field $\mathbf{l} = \mathbf{k}(b_1)$ for t_2 is generated by a square root external to the original splitting field $\mathbf{K}$—an **auxiliary irrationality**. (Exercise 5.5.1 will show this.) In contrast, the square root of the discriminant, which lies within $\mathbf{K}$, is a **natural irrationality**. Using an auxiliary irrationality to simplify p, i.e., leaving its splitting field, feels inelegant and clumsy, and raises the question of whether p can be comparably simplified without auxiliary quantities.

How are $\mathrm{spl}_\mathbf{l}(q)$ and $\mathrm{Gal}_\mathbf{l}(q)$ related to $\mathbf{L} = \mathrm{spl}_\mathbf{l}(p)$ and $\Gamma = \mathrm{Gal}_\mathbf{l}(p)$; that is, how is solving q over $\mathbf{l}$ related to solving p over $\mathbf{l}$? And how is Γ related to $G = \mathrm{Gal}_\mathbf{k}(p)$; that is, how is solving p over $\mathbf{l}$ related to solving p over $\mathbf{k}$? The next section will show that generally $\mathrm{spl}_\mathbf{l}(q) = \mathbf{L}$ so that $\mathrm{Gal}_\mathbf{l}(q) = \Gamma$, i.e., solving q solves p once the coefficient field $\mathbf{l}$ is constructed. This is as we would hope. Less happily, usually $\Gamma = G$, meaning a Tschirnhaus transformation typically does not diminish the Galois group of a polynomial p even by introducing auxiliary quantities.

While the auxiliary Tschirnhaus coefficients and persistent Galois group are discouraging, the Tschirnhaus transformation works wonders at diminishing parameter-count. The transformations t_1 and t_2 combine to knock out two coefficients of the general polynomial p, and further Tschirnhaus transformation will reduce the three-parameter principal quintic to one-parameter Brioschi form without any more auxiliary irrationalities. Chapter 6 will prove that any reduction of the quintic by radicals to one-parameter form requires at least one auxiliary irrationality, so the Brioschi reduction, with its single quadratic auxiliary b_1, is optimal.

Exercises

5.4.1. (a) Verify that $q = cR(p, t - S)$ for some constant $c \in \mathbf{l}$ by arguing that the resultant is indeed a polynomial of degree $\deg(p)$ in S, and it vanishes at the same values of S as q.

(b) The resultant in (a) is a determinant of order $n{\cdot}\tau$ where $n = \deg(p)$ and $\tau = \deg(t)$. Here is a smaller determinant (of order n) that also gives q. Show that if $p(T) = 0$ and $S = t(T)$ (where $p \in \mathbf{k}[T]$ has degree n and $t \in \mathbf{l}[T]$ has smaller degree) then there exist polynomials

$t_1, t_2, \ldots, t_{n-1} \in \mathbf{l}[T]$, each of degree less than n, such that

$$S = t(T), \quad ST = t_1(T), \quad ST^2 = t_2(T), \quad \ldots, \quad ST^{n-1} = t_{n-1}(T).$$

Equivalently,

$$S \cdot \begin{bmatrix} 1 \\ T \\ \vdots \\ T^{n-1} \end{bmatrix} = M_t \cdot \begin{bmatrix} 1 \\ T \\ \vdots \\ T^{n-1} \end{bmatrix}$$

where M_t is an n-by-n matrix over $\mathbf{l}$. This shows that $\det(M_t - SI) = 0$, i.e., q is the characteristic polynomial of M_t.

5.4.2. Confirm the computation of the affine Tschirnhaus transformation t_1 that reduces the general polynomial to depressed form.

5.4.3. This exercise computes the quadratic Tschirnhaus transformation that reduces the depressed polynomial of degree n to principal form. Given

$$p = T^n + a_2 T^{n-2} + \cdots + a_n = \prod_{i=1}^{n} (T - r_i)$$

with $a_2 \neq 0$, the idea is to choose the transformation $t_2 = T^2 + b_1 T + b_2$ so that the resulting polynomial

$$q = \prod_{i=1}^{n} (T - t_2(r_i)) = T^n + c_1 T^{n-1} + \cdots + c_n$$

has first two coefficients $c_1 = c_2 = 0$. Show that $c_1 = -s_2 - nb_2$ where $s_2 = s_2(r_1, \ldots, r_n)$ is the second power sum of the roots of p, so necessarily $b_2 = -s_2/n$. By Newton's identities, it now suffices (Proposition 5.2.1) to make $s_2(t_2(r_1), \ldots, t_2(r_n)) = 0$. Show that this is equivalent to b_1 satisfying the quadratic polynomial

$$Q = s_2 T^2 + 2s_3 T + s_4 - s_2^2/n \in \mathbf{k}[T],$$

where the s_j are the power sums of $r_1, \ldots, r_n$. Solving this quadratic for b_1 gives the desired transformation.

5.4.4. Specializing the previous exercise to the case $n = 3$, show that the quotient of discriminants $\Delta(p)/\Delta(Q)$ is a square in $\mathbf{C}(\sigma_1, \sigma_2, \sigma_3)$, so adjoining the square root of either discriminant generates the same field. (A symbolic algebra computer package may speed this up.) Thus t_2 solves the general depressed cubic by reducing its Galois group to A_3, and it does so without introducing an auxiliary irrationality.

5.4.5. Use a computer algebra system to implement the transformations t_1, t_2 of this section and work examples. (Recall that the transformed polynomial is a resultant, which the algebra system will know how to calculate. For t_2, use Newton's identities to compute the power sums s_2, s_3, s_4 from the coefficients of p.)

5. Galois theory of the Tschirnhaus transformation

Suppose that $\mathbf{k}$ is the coefficient field of a polynomial p with splitting field $\mathbf{K} = \mathrm{spl}_{\mathbf{k}}(p) = \mathbf{k}(r_1, \ldots, r_n)$ where the r_i are the roots of p. Let $G = \mathrm{Gal}_{\mathbf{k}}(p)$. Let $\mathbf{l}$ be any extension of $\mathbf{k}$. Let $\mathbf{L} = \mathrm{spl}_{\mathbf{l}}(p) = \mathbf{l}(r_1, \ldots, r_n) = \mathbf{Kl}$ and $\Gamma = \mathrm{Gal}_{\mathbf{l}}(p)$. See Figure 5.4.1 again for the diagram.

Any automorphism $\sigma \in \Gamma$ fixes $\mathbf{k}$ and therefore permutes the roots r_i of the polynomial $p \in \mathbf{k}[T]$, so it restricts to an automorphism of $\mathbf{K}$ over $\mathbf{k}$, i.e., an element of G. The next result shows that distinct elements of Γ restrict to distinct elements of G, i.e., Γ naturally embeds in G.

(5.5.1) PROPOSITION. *In the situation above, the restriction map from Γ to G is injective. Its image is Gal$(\mathbf{K}/\mathbf{l} \cap \mathbf{K})$.*

PROOF. If the restriction of some $\sigma \in \Gamma$ fixes $\mathbf{K}$ along with $\mathbf{l}$ then it fixes $\mathbf{Kl} = \mathbf{L}$ and is trivial. This proves the first statement.

For the second assertion, now identifying Γ with its natural image in G, it suffices to show that $\mathbf{K}^\Gamma = \mathbf{l} \cap \mathbf{K}$. Clearly $\mathbf{l} \cap \mathbf{K} \subset \mathbf{K}^\Gamma$. And if $x \in \mathbf{K}^\Gamma$ then $x \in \mathbf{L}^\Gamma = \mathbf{l}$, so $x \in \mathbf{l} \cap \mathbf{K}$, giving the other containment. □

For notational convenience, identify Γ with its natural image $\mathrm{Gal}(\mathbf{K}/\mathbf{l} \cap \mathbf{K}) \subset G$ from here on.

Now take a Tschirnhaus transformation $t = b_0 T^k + b_1 T^{k-1} + \cdots + b_k$, specialize the discussion to $\mathbf{l} = \mathbf{k}(b_0, \ldots, b_k)$, and set $q = \prod_{i=1}^n (S - t(r_i)) \in \mathbf{l}[S]$. Assume that if roots r_i, r_j of p are distinct then so are $t(r_i)$ and $t(r_j)$, i.e., t doesn't collapse the roots. Call such a Tschirnhaus transformation **nondegenerate**. For such t, $\mathrm{spl}_{\mathbf{l}}(q) = \mathbf{L}$ and $\mathrm{Gal}_{\mathbf{l}}(q) = \Gamma$. To see this, inspect the fields and groups in Figure 5.5.1, where all extensions are Galois. Since any permutation of $\{r_1, \ldots, r_n\}$ induces the corresponding permutation of $\{t(r_1), \ldots, t(r_n)\}$, H is trivial, giving the result.

This discussion proves

(5.5.2) THEOREM. *Let p, t and q be as above. Let $\mathbf{k}$ be the coefficient field of p, $\mathbf{l}$ the coefficient field of q, $\mathbf{K} = \mathrm{spl}_{\mathbf{k}}(p)$ and $\mathbf{L} = \mathbf{Kl} = \mathrm{spl}_{\mathbf{l}}(p)$. Let $G = Gal_{\mathbf{k}}(p)$ and $\Gamma = Gal_{\mathbf{l}}(p)$. Then $\Gamma \subset G$ and $[G : \Gamma] = [\mathbf{l} \cap \mathbf{K} : \mathbf{k}]$. If t is nondegenerate then $\mathbf{L} = \mathrm{spl}_{\mathbf{l}}(q)$ and $\Gamma = Gal_{\mathbf{l}}(q)$.*

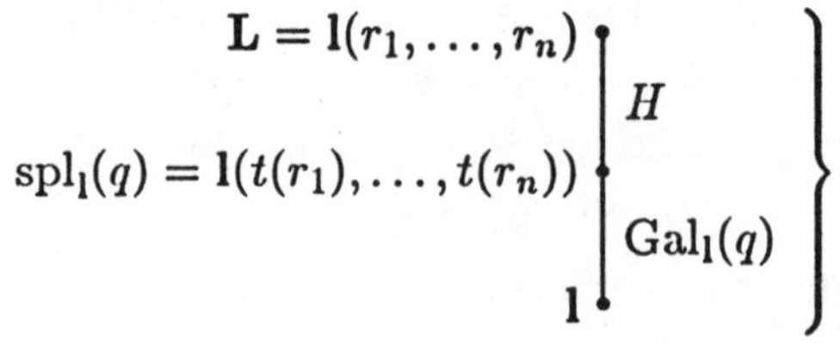

$$\left.\begin{array}{rl} \mathbf{L} = \mathbf{l}(r_1, \ldots, r_n) & \\ & \Big| \; H \\ \mathrm{spl}_{\mathbf{l}}(q) = \mathbf{l}(t(r_1), \ldots, t(r_n)) & \\ & \Big| \; \mathrm{Gal}_{\mathbf{l}}(q) \\ \mathbf{l} & \end{array}\right\} \Gamma$$

Figure 5.5.1.

In particular, nondegenerate Tschirnhaus transformations with $\mathbf{l} = \mathbf{k}$ are Galois-theoretically trivial over $\mathbf{k}$: the splitting field and Galois group are left unaltered by such a transformation from p to q. We saw this for the affine t_1 that depressed the general polynomial in the preceding section.

For the quadratic transformation t_2 that reduced the depressed polynomial to principal form, we have $\mathbf{l} = \mathbf{k}(b_1)$ where b_1 satisfies the quadratic Q in Exercise 5.4.3. There are three possibilities:

1. If $\Delta(Q)$ is a square in $\mathbf{k}$ then $\mathbf{l} = \mathbf{k}$ and $\Gamma = G$.
2. If $\Delta(Q)$ is a square in $\mathbf{K}$ but not in $\mathbf{k}$ then $\mathbf{k} \subset \mathbf{l} \subset \mathbf{K}$ and $[G : \Gamma] = 2$, so t_2 reduces the Galois group of p without leaving the splitting field. Exercise 5.4.4 showed that this happens for the general depressed p when $n = 3$.
3. If $\Delta(Q)$ is not a square in $\mathbf{K}$ then $\mathbf{l}$ is external to $\mathbf{K}$ and $\Gamma = G$. Exercise 5.5.1 shows that this happens for the general depressed p when $n > 3$; Exercise 5.5.2 shows that for such p and n, t_2 is nondegenerate. Thus for $n > 3$, reducing the general polynomial to principal form introduces an auxiliary irrationality; since $\mathrm{spl}_{\mathbf{l}}(q) = \mathrm{spl}_{\mathbf{l}}(p)$, solving q over $\mathbf{l}$ solves p, but is Galois-theoretically no easier than the original problem of solving p over $\mathbf{k}$.

Exercises

5.5.1. Let p be the general depressed polynomial p of degree n, whose roots sum to zero. Its coefficient field is thus $\mathbf{k} = \mathbf{C}(\sigma_2, \ldots, \sigma_n)$ and its Galois group is S_n. This exercise shows that when $n > 3$, applying t_2 to reduce p to principal form introduces an auxiliary irrationality, in contrast to the cubic case.

(a) Let Q be the quadratic polynomial from Exercise 5.4.3. Calculate $\Delta(Q)$ and show that it has homogeneous degree 6 in the roots of p. Show that $\Delta(Q)$ is not a square in $\mathbf{k}$, so that the root b_1 of Q is external

to **k**. (Hint: prove the lemma that for any unique factorization domain R with field of quotients F, if $s \in R$ is a square in F then s is a square in R. Now let R be the polynomial ring $\mathbf{C}[\sigma_2, \ldots, \sigma_n]$ (you may cite that this is a unique factorization domain) with field of quotients **k**. By the lemma, a proof by contradiction begins by assuming $\Delta(Q) = g^2$ for some $g \in R$.) Note that this argument works for all $n \geq 3$.
(b) Show that $\Delta(p)$ is irreducible in $R = \mathbf{C}[\sigma_2, \ldots, \sigma_n]$. (Recall how $\Delta(p)$ factors in $\mathbf{C}[r_1, \ldots, r_n]$ and show that no proper factor of $\Delta(p)$ is invariant under S_n.) Show that $\Delta(p)$ has homogeneous degree $n(n-1)$ in the roots of p. Since $6 < n(n-1)$ for all $n > 3$, $\Delta(Q)$ and $\Delta(p)$ therefore are relatively prime in R. Show that $\Delta(p)/\Delta(Q)$ is not a square in **k**. (Hint: prove the lemma that for a, b, c, d in a unique factorization domain R with field of quotients F, if $a/b = c/d$ with $\gcd(a,b) = \gcd(c,d) = 1$ then $c = ua$ and $d = ub$ for some unit $u \in R^*$. Now do a proof by contradiction.) It follows that for $n > 3$, the quadratic extension $\mathbf{k}(b_1)$ is distinct from $\mathbf{k}(\sqrt{\Delta(p)})$, the unique quadratic extension of **k** in **K**.
(c) Strictly speaking, what we really want to know is whether applying $t_2 \circ t_1$ to the original general polynomial of degree $n > 3$ introduces an auxiliary irrationality over the original coefficient field. Show that in fact it does. (Hint: rewrite Q in terms of the power sums of the original polynomial and proceed as in parts (a) and (b).)

5.5.2. Show that the quadratic transformation t_2 is nondegenerate unless $b_1 = -r_i - r_j$ for some distinct r_i, r_j. Since $-r_i - r_j$ lies in **K**, this shows that t_2 is nondegenerate for the general depressed polynomial when $n > 3$. Thus, solving the general principal polynomial over **l** solves the general polynomial over **k**.

6. Projective space and algebraic sets

While one more Tschirnhaus transformation reduces the quintic from principal to one-parameter Brioschi form, and this transformation can be described purely algebraically, we first develop some geometry in the next two sections indicating how the principal quintic is related to the rotations of the icosahedron, which in turn (see Exercise 4.8.6) are related to a family of principal quintics. While the geometry is appealing, it still takes considerable work to solve the general principal quintic by icosahedral methods, and since the methods involve three parameters rather than one, we omit them. The relevant calculations are in Klein and Dickson, and Klein motivates them geometrically. Reduction of the quintic to Brioschi form resumes in Section 5.8, also guided by this geometry.

For any positive integer n, **n-dimensional complex projective space** is defined analogously to the complex projective line,

$$\mathbf{P}^n(\mathbf{C}) = (\mathbf{C}^{n+1} \setminus \{\mathbf{0}\})/\sim,$$

where $x \sim y$ in $\mathbf{C}^{n+1}$ if $y = \lambda x$ for some $\lambda \in \mathbf{C}^*$. The image of a point $z = (z_1, \ldots, z_{n+1})$ under the natural map $\mathbf{C}^{n+1} \setminus \{\mathbf{0}\} \longrightarrow \mathbf{P}^n(\mathbf{C})$ is written $p = [z_1 : \cdots : z_{n+1}]$. (Throughout this section the symbol p denotes a point in $\mathbf{P}^n(\mathbf{C})$ rather than a polynomial.) The coordinates of p are determined only up to a common scalar multiple. The vector space structure of $(n+1)$-dimensional **affine space** $\mathbf{C}^{n+1}$ does not transfer to $\mathbf{P}^n(\mathbf{C})$—try to define addition and scalar multiplication to see this. Projective space does inherit the quotient topology from $\mathbf{C}^{n+1} \setminus \{\mathbf{0}\}$. It is compact, being the continuous image of the sphere $\{(z_1, \ldots, z_{n+1}) \in \mathbf{C}^{n+1} : \sum_{i=1}^{n+1} |z_i|^2 = 1\}$, which is closed and bounded in $\mathbf{R}^{2n+2}$.

Since $\mathbf{P}^n(\mathbf{C}) = [\mathbf{C}^n : 1] \cup [\mathbf{P}^{n-1}(\mathbf{C}) : 0]$, projective space $\mathbf{P}^n(\mathbf{C})$ contains a copy of n-dimensional affine space $\mathbf{C}^n$ along with some lower-dimensional projective material. A similar decomposition can be carried out by setting any coordinate, not necessarily the last one, to 1 and 0. In fact projective coordinates are also called **homogeneous** coordinates to emphasize that they all enjoy equal status, unlike, for example, the numerator and denominator of a fraction. This was the motivation for identifying the Riemann sphere $\widehat{\mathbf{C}}$ with the complex projective line $\mathbf{P}^1(\mathbf{C})$ in Chapter 1 to avoid divide-by-zero problems. Define for $i = 1, \ldots, n+1$ the set $U_i = \{p \in \mathbf{P}^n(\mathbf{C}) : z_i \neq 0\}$. Each local coordinate map

$$c_i : U_i \longrightarrow \mathbf{C}^n \quad \text{given by} \quad c_i[\cdots : z_{i-1} : 1 : z_{i+1} : \cdots] = (\ldots, z_{i-1}, z_{i+1}, \ldots)$$

is a homeomorphism, cf. the discussion of the projective line in Section 1.4. Since $\mathbf{P}^n(\mathbf{C}) = \cup U_i$, projective space is covered by $n+1$ copies of affine space.

An **algebraic set** in $\mathbf{P}^n(\mathbf{C})$ is the set of common zeros of a finite set $\mathcal{F}$ of forms in $\mathbf{C}[Z_1 : \cdots : Z_{n+1}]$,

$$\mathcal{Z}(\mathcal{F}) = \{p \in \mathbf{P}^n(\mathbf{C}) : f(p) = 0 \text{ for all } f \in \mathcal{F}\}.$$

For example, the zero set of $\mathcal{F} = \{X^2+Y^2-Z^2\}$ in $\mathbf{P}^2(\mathbf{C})$ is the affine "unit circle" $[\mathcal{Z}(X^2+Y^2-1) : 1]$ and the points $[1 : \pm i : 0]$, while the zero set of $\mathcal{G} = \{X^2+Y^2-Z^2, Y\}$ in $\mathbf{P}^2(\mathbf{C})$ is $\{[\pm 1 : 0 : 1]\}$. A **hypersurface** is the zero-set of one form; its **degree** is the degree of the form. A **hyperplane** is a hypersurface of degree 1, i.e., the zero-set of one linear form $\sum_{i=1}^{n+1} a_i Z_i$. The **tangent hyperplane** to a hypersurface $S = \mathcal{Z}(f)$ at the point $p \in S$, denoted $T_p(S)$, is the hyperplane $\mathcal{Z}(\sum_{i=1}^{n+1} D_i f(p) Z_i)$, provided at least one partial derivative $D_i f(p)$ is nonzero. This is well-defined, and Euler's identity shows that indeed the point p lies in $T_p(S)$. (See Exercise 5.6.2.) A **line** in $\mathbf{P}^n(\mathbf{C})$ is the intersection of $n-1$ hyperplanes, $L = \mathcal{Z}(\{\sum_{j=1}^{n+1} a_{ij} Z_j : 1 \le i \le n-1\})$, where the coefficient matrix $[a_{ij}] \in \mathrm{M}_{n-1,n+1}(\mathbf{C})$ has full rank. Exercise 5.6.4 shows that the line through points $p, q \in \mathbf{P}^n(\mathbf{C})$ also takes the parametrized (and well-defined) form $L_{p,q} = \{\lambda_1 p + \lambda_2 q : [\lambda_1 : \lambda_2] \in \mathbf{P}^1(\mathbf{C})\}$. A **quadric** is a hypersurface of degree 2. One can check that when the underlying field $\mathbf{C}$ is replaced by $\mathbf{R}$ on each U_i, the surfaces, hyperplanes, etc. described here are indeed the appropriate affine objects. See Exercises 5.6.3 and 5.6.5.

Each automorphism (invertible linear self-map) L of $\mathbf{C}^{n+1}$ induces a corresponding **projective transformation** P of $\mathbf{P}^n(\mathbf{C})$, defined to make the following diagram commute.

$$\begin{array}{ccc} \mathbf{C}^{n+1} \setminus \{\mathbf{0}\} & \xrightarrow{L} & \mathbf{C}^{n+1} \setminus \{\mathbf{0}\} \\ \downarrow & & \downarrow \\ \mathbf{P}^n(\mathbf{C}) & \xrightarrow{P} & \mathbf{P}^n(\mathbf{C}) \end{array}$$

The natural identification of the automorphisms of $\mathbf{C}^{n+1}$ with the matrix group $\mathrm{GL}_{n+1}(\mathbf{C})$ identifies the projective transformations of $\mathbf{P}^n(\mathbf{C})$ with the projective matrix group $\mathrm{PGL}_{n+1}(\mathbf{C}) = \mathrm{GL}_{n+1}(\mathbf{C})/\mathbf{C}^* I$. Every permutation $\sigma \in S_{n+1}$ induces an automorphism of $\mathbf{C}^{n+1}$ and therefore a projective transformation of $\mathbf{P}^n(\mathbf{C})$. Every projective transformation takes hypersurfaces to hypersurfaces of the same degree and is a **collineation**, meaning it takes lines to lines. Exercise 5.6.6 asks for proofs of these statements.

Exercises

5.6.1. Prove that the local coordinate maps $c_i : U_i \longrightarrow \mathbf{C}^n$ are homeomorphisms.

5.6.2. Let $S = \mathcal{Z}(f) \subset \mathbf{P}^n(\mathbf{C})$ be a hypersurface and let p be a point in S. Assume that some partial derivative $D_i f(p)$ is nonzero. Show that the tangent hyperplane $T_p(S) = \mathcal{Z}(\sum_{i=1}^{n+1} D_i f(p) Z_i)$ is independent of the choice of coordinates for p. Prove Euler's identity: for any form $f \in \mathbf{C}[Z_1 : \cdots : Z_{n+1}]$, $\sum_{i=1}^{n+1} D_i f\, Z_i = \deg(f) \cdot f$. How does this show that $p \in T_p(S)$?

5.6.3. Let $S = \mathcal{Z}(f) \subset \mathbf{P}^n(\mathbf{C})$ be a hypersurface and let p be a point in S. Suppose that $p \in U_{n+1}$, i.e., $p = [p_1 : \cdots : p_n : 1]$, and that some $D_i f(p)$ is nonzero. Show that $S \cap U_{n+1} = [\mathcal{Z}(f_*) : 1]$ where f_* is the dehomogenized polynomial $f_*(Z_1, \ldots, Z_n) = f(Z_1, \ldots, Z_n, 1)$. Show that $T_p(S) \cap U_{n+1}$ is defined by the condition

$$\sum_{i=1}^{n} D_i f_*(p_1, \ldots, p_n)(Z_i - p_i) = 0$$

which describes the affine tangent hyperplane to $S \cap U_{n+1}$ at p. (You may need Euler's identity.)

5.6.4. (a) For any distinct points $p, q \in \mathbf{P}^n(\mathbf{C})$ show that the set $L_{p,q} = \{\lambda_1 p + \lambda_2 q : [\lambda_1 : \lambda_2] \in \mathbf{P}^1(\mathbf{C})\}$ is well-defined, i.e., is independent of the choice of coordinates for p and q.

(b) Let $L \subset \mathbf{P}^n(\mathbf{C})$ be a line, i.e., the projective image of the nonzero solutions in $\mathbf{C}^{n+1}$ to the system $AZ = 0$ where $A \in \mathrm{M}_{n-1,n+1}\mathbf{C}$ has full rank. Show that there exist distinct points $p, q \in \mathbf{P}^n(\mathbf{C})$ such that $L = L_{p,q}$. (Hint: rank–nullity theorem.) Conversely, given distinct $p, q \in \mathbf{P}^n(\mathbf{C})$, show that $L_{p,q}$ is indeed a line.

5.6.5. Let L be a line in $\mathbf{P}^n(\mathbf{C})$. Show that $L \cap U_{n+1}$ is empty or takes the form $[L' : 1]$ where L' is a line in affine space.

5.6.6. Verify that projective transformations of $\mathbf{P}^n(\mathbf{C})$ are well-defined and naturally identified with $\mathrm{PGL}_{n+1}(\mathbf{C})$. Explain how any permutation $\sigma \in S_{n+1}$ induces an automorphism of $\mathbf{C}^{n+1}$ and therefore a projective transformation of $\mathbf{P}^n(\mathbf{C})$. Show that every projective transformation takes hypersurfaces to hypersurfaces of the same degree and is a collineation.

7. Geometry of the Tschirnhaus transformation

Returning to the polynomial $p = \prod_{i=1}^{n}(T - r_i)$ of degree n, think now of the roots r_i as complex variables rather than algebraic symbols. Once the roots have been indexed, any Tschirnhaus transformation t can be viewed geometrically as moving the root vector $(r_1, \ldots, r_n) \in \mathbf{C}^n$ to $(t(r_1), \ldots, t(r_n))$. Affine Tschirnhaus transformations $t = b_0 T + b_1$ (with $b_0 \neq 0$) are invertible: if such t transforms the polynomial p to $q = \prod(S - t(r_i))$ then $t^{-1} = b_0^{-1}(S - b_1)$ transforms q back to p. Thus from the viewpoint of Tschirnhaus transformations, translating root vectors (specifically, by the transformation t_1 in Section 5.4 to make their entries sum to zero) and scaling root vectors are trivial operations, and it is consequently natural to study

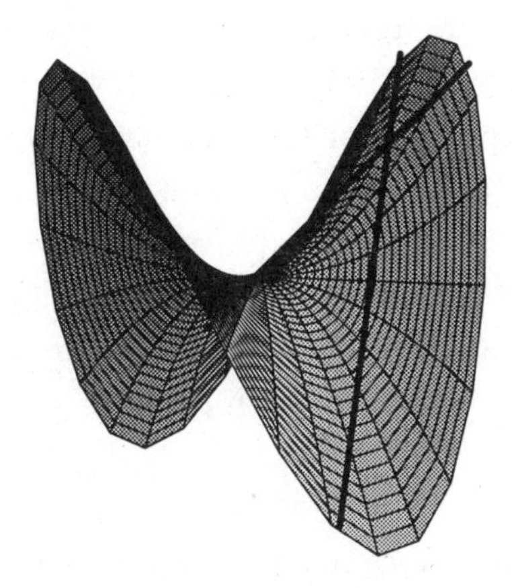

Figure 5.7.1. Lines on the quadric surface $XY = ZW$

Tschirnhaus transformations as acting on the hyperplane $H = \mathcal{Z}(\sum_{i=1}^{n} Z_i)$ of $\mathbf{P}^{n-1}(\mathbf{C})$. This effectively reduces the root space dimension by 2. Indeed, we may as well identify H with $\mathbf{P}^{n-2}(\mathbf{C})$ via the collineation $[z_1 : \cdots : z_{n-1} : -\sum_{i=1}^{n-1} z_i] \mapsto [z_1 : \cdots : z_{n-1}]$.

Newton's identities show (Proposition 5.2.1) that reducing a polynomial to principal form (by the transformations t_1 and t_2) is equivalent to transforming its root vector $(r_1, \ldots, r_n)$ onto the **canonical surface** $C = \mathcal{Z}(\sum_{i=1}^{n} Z_i, \sum_{i=1}^{n} Z_i^2)$. The identification of H with $\mathbf{P}^{n-2}(\mathbf{C})$ maps the canonical surface to the quadric $Q = \mathcal{Z}(\sum_{i=1}^{n-1} Z_i^2 + (\sum_{i=1}^{n-1} Z_i)^2) \subset \mathbf{P}^{n-2}(\mathbf{C})$. Exercises 5.7.1 and 5.7.2 combine to show that when $n = 5$, i.e., p is quintic, the Tschirnhaus transformation that reduces p to principal form moves its roots onto a quadric surface collinear to $\mathcal{Z}(XY - ZW) \subset \mathbf{P}^3(\mathbf{C})$.

Exercise 5.7.3 shows that $\mathcal{Z}(XY - ZW)$ consists of two one-parameter families of lines: $\{L_\lambda : \lambda \in \widehat{\mathbf{C}}\}$ and $\{L'_\mu : \mu \in \widehat{\mathbf{C}}\}$. (See Figure 5.7.1.) These lines give the connection between the principal quintic and icosahedral geometry; here is the idea, without proof. Every permutation $\sigma \in S_5$ defines a corresponding collineation of projective space $\mathbf{P}^4(\mathbf{C})$ (Exercise 5.6.6). These preserve the canonical surface and hence correspond to collineations of $\mathcal{Z}(XY - ZW)$. It turns out that the odd permutations $\sigma \in S_5$ exchange the two families of lines—i.e., an odd σ induces a collineation taking each L_λ to some $L'_{\sigma(\lambda)}$ and each L'_μ to some $L_{\sigma(\mu)}$—while the even permutations σ preserve the families, taking each L_λ to some $L_{\sigma(\lambda)}$ and similarly for the $\{L'\}$. Thus, given the general principal quintic, adjoin the square root of its discriminant to its coefficient field; the resulting Galois group A_5 acts twice on the Riemann sphere by each $\sigma \in A_5$ taking $\lambda \mapsto \sigma(\lambda)$ as just described, and similarly for μ. The action is faithful. Now, by Theorem 2.3.1, each A_5-induced automorphism group of $\widehat{\mathbf{C}}$ is conjugate to a rotation group, and by Theorem 2.6.1, the rotation group must be icosahedral. Thus the coefficients of a principal quintic are icosahedral invariants of the line coordinates (λ, μ) of its root vector, and plausibly the general principal quintic, with its three independent coefficients, is equivalent to the principal quintic icosahedral resolvent $R_{\tilde{u}}$ from Exercise 4.8.6, with its three parameters d_1, d_2, and W'.

Exercises

5.7.1. This exercise describes quadric hypersurfaces in $\mathbf{P}^n(\mathbf{C})$ modulo projective transformations.

(a) Show that any quadric hypersurface takes the form

$$Q = \{[z_1 : \cdots : z_{n+1}] : z^t A z = 0\}$$

where $A \in \mathrm{M}_{n+1}(\mathbf{C})$ is symmetric and defined only up to scalar multiple.

(b) Show that any projective transformation $P^{-1} \in \mathrm{PGL}_{n+1}(\mathbf{C})$ transforms Q to the surface defined by $P^t A P$. Thus, classifying quadrics modulo change of variable is equivalent to classifying matrix **congruence classes** $\{P^t A P : P \in \mathrm{PGL}_{n+1}(\mathbf{C})\}$. (Recall an analogous situation from linear algebra: classifying linear transformations of $\mathbf{C}^n$ up to change of coordinates amounts to classifying matrix equivalence classes $\{P^{-1} A P : P \in \mathrm{GL}_n(\mathbf{C})\}$.)

(c) Prove that any symmetric square matrix $A \in \mathrm{M}_{n+1}(\mathbf{C})$ is congruent to the matrix $\begin{bmatrix} I_r & 0 \\ 0 & 0 \end{bmatrix}$ where r is the rank of A, as follows. There exist vectors $v, w \in \mathbf{C}^{n+1}$ such that $v^t A w \neq 0$ unless A is the already-diagonal zero matrix. At least one of the following must hold: $v^t A v \neq 0$, $w^t A w \neq 0$, or $(v+w)^t A (v+w) \neq 0$. Thus there exists a vector u such that $u^t A u = 1$. Define the linear map $\mathbf{C}^{n+1} \longrightarrow \mathbf{C}$ by $x \mapsto u^t A x$. This surjects, so its kernel has dimension n. Let $\{u_1, \ldots, u_n\}$ be a basis for the kernel and form the matrix $P = [u \quad u_1 \quad \cdots \quad u_n]$. Then P is invertible, and

$$P^t A P = \begin{bmatrix} 1 & 0_{1\times n} \\ 0_{n\times 1} & A'_{n\times n} \end{bmatrix}$$

where A' is also symmetric. The proof is completed by induction. This shows that any quadric surface in $\mathbf{P}^n(\mathbf{C})$ is equivalent under projective transformation to a surface $\mathcal{Z}(\sum_{i=1}^{r} Z_i^2)$.

(d) Prove that the 2-by-2 identity matrix is congruent over $\mathbf{C}$ to either of $\pm\frac{1}{2}\begin{bmatrix} 0 & 1 \\ 1 & 0 \end{bmatrix}$. With part (c) this shows that any full-rank quadric surface in $\mathbf{P}^3(\mathbf{C})$ is equivalent under projective transformation to the surface $\mathcal{Z}(XY - ZW)$, which is often called **the** quadric surface in $\mathbf{P}^3(\mathbf{C})$.

5.7.2. Show that the matrix defining the quadric Q in the section is

$$A = \begin{bmatrix} 2 & 1 & \cdots & 1 \\ 1 & 2 & \cdots & 1 \\ \vdots & \vdots & \ddots & \vdots \\ 1 & 1 & \cdots & 2 \end{bmatrix}.$$

Show that A's rows are linearly independent.

5.7.3. This exercise shows that the quadric surface $Q = \mathcal{Z}(XY - ZW)$ in $\mathbf{P}^3(\mathbf{C})$ is doubly ruled, i.e., there exist two $\mathbf{P}^1(\mathbf{C})$-parametrized families of lines each of which forms Q. The previous exercise then implies that all full-rank quadric surfaces in $\mathbf{P}^3(\mathbf{C})$ are doubly ruled.

(a) Informally, the lines are $L_\lambda = \{[x:y:z:w] : x/z = w/y = \lambda\}$ and $L'_\mu = \{[x:y:z:w] : x/w = z/y = \mu\}$ for $\lambda, \mu \in \widehat{\mathbf{C}}$. To make this precise, show that the map $\mathbf{P}^1(\mathbf{C}) \times \mathbf{P}^1(\mathbf{C}) \longrightarrow Q$ where

$$([\lambda_1 : \lambda_2], [\mu_1 : \mu_2]) \mapsto [\lambda_1\mu_1 : \lambda_2\mu_2 : \lambda_2\mu_1 : \lambda_1\mu_2]$$

is a bijection by exhibiting its inverse. Show that each $\{\lambda\} \times \mathbf{P}^1(\mathbf{C})$ maps to a line L_λ in Q and similarly for each $\mathbf{P}^1(\mathbf{C}) \times \{\mu\}$. Thus $Q = \cup_\lambda L_\lambda = \cup_\mu L'_\mu$, and both unions are disjoint.

(b) Prove that part (a) exhibits all lines in Q by showing that for any $p = [a:b:c:d] \in Q$, $Q \cap T_p(Q)$ consists of two lines. Take $d = 1$ without loss of generality, so $ab = c$; then points $[x:y:z:w] \in Q \cap T_p(Q)$ satisfy

$$xy - zw = 0, \qquad bx + ay - z - abw = 0.$$

Substituting the second condition $z = bx + ay - abw$ into the first gives $(x - aw)(y - bw) = 0$, indeed defining the two lines L'_a and $L_{b^{-1}}$.

8. Brioschi form

Having reduced the general quintic to principal form, we may apply the general notation to the principal case. Thus the general principal quintic is now

$$p = T^5 - \sigma_3 T^2 + \sigma_4 T - \sigma_5 = \prod_{i=1}^{5}(T - r_i).$$

The field generated by the coefficients of p and the square root of its discriminant is $\mathbf{k} = \mathbf{C}(\sigma_3, \sigma_4, \sigma_5, \sqrt{\Delta(p)})$. This field includes the linear coefficient b_1 of the Tschirnhaus transformation t_2 that reduced the depressed quintic to principal form, and this b_1 is an auxiliary square root over the coefficient field of the original general quintic, but that is irrelevant to this section, which starts from the principal case. The square root $\sqrt{\Delta(p)}$, by contrast, is a natural irrationality over the coefficient field of the principal quintic p since it lies in the splitting field. The principal quintic has three parameters σ_3, σ_4, σ_5; the goal is to transform it to a one-parameter form. The hard work is reducing to two parameters, after which an easy scaling finishes the job.

Chapter II.1 of Klein gives the history of this reduction. The original insight came from Jacobi in 1829, arising from so-called elliptic modular functions. Kronecker and Brioschi expanded the idea, computing from the general quintic a sextic resolvent that in turn has a quintic resolvent of the desired form. This two-step calculation was then refined into a purely algebraic one-step procedure, given here with some geometrical motivation. Since the algebra is fairly involved, a numerical example will follow at the end of the section.

Recall that the root vector $(r_1, \ldots, r_5)$ of p satisfies

$$\sum r_i = \sum r_i^2 = 0.$$

(All sums in this section are for $i = 1$ to 5.) As explained in the previous section, this means that each principal quintic in $\mathbf{C}[T]$ with numerical root vector $(r_1, \ldots, r_5) \in \mathbf{C}^5$ describes a point on the canonical surface C.

(5.8.1) PROPOSITION. *There exists a polynomial $\psi \in \mathbf{k}[T]$, monic of degree 3 or 4, such that*

$$\sum \psi(r_i) = \sum \psi(r_i)^2 = \sum r_i \psi(r_i) = 0.$$

PROOF. Exercise 5.8.1. Constructing ψ is straightforward but uses a square root; the surprising fact is that this square root already lies in $\mathbf{k}$, so the construction requires no irrationalities. □

(5.8.2) COROLLARY. *Let $(r_1, \ldots, r_5) \in \mathbf{C}^5$ lie on the canonical surface C. For any $a, b \in \mathbf{C}$, the $(aT + b\psi)$-transformed vector*

$$(ar_1 + b\psi(r_1), \ldots, ar_5 + b\psi(r_5))$$

also lies on C. Equivalently, the $(aT+b\psi)$-transform of the principal quintic p,

$$q = \prod_{i=1}^{5}(T - (ar_i + b\psi(r_i))),$$

is also principal.

PROOF. Exercise 5.8.2. □

The plane's worth of transformed vectors in Corollary 5.8.2 corresponds to a projective line through $[r_1 : \cdots : r_5]$ in the canonical surface; for each principal quintic in $\mathbf{C}[T]$, a choice of square root in Exercise 5.8.1 determines which of the two possible lines is obtained. Thus, the general process of finding the lines through a point in C takes place in $\mathbf{k}$.

Next recall that any Tschirnhaus transformation taking roots r_i of p to new roots $t(r_i)$, where t is a polynomial, only depends on t mod p. The next result provides a congruence mod p and therefore exhibits two polynomials that transform p in the same way.

(5.8.3) PROPOSITION. *There exist a linear form and a quadratic form in $\mathbf{k}[X, Y]$,*

$$L(X,Y) = aX + bY \qquad \textit{and} \qquad Q(X,Y) = \alpha X^2 + 2\beta XY + \gamma Y^2,$$

such that, if ψ is as in Proposition 5.8.1,

$$L(T,\psi) \equiv Q(T,\psi) \pmod{p}.$$

(Note that the congruence is in $\mathbf{k}[T]$ since $\psi \in \mathbf{k}[T]$.)

PROOF. Exercise 5.8.3. □

Thus, as remarked before the proposition, $L(T,\psi)$ and $Q(T,\psi)$ act as the same Tschirnhaus transformation on the roots of the principal quintic. Transforming by $L(T,\psi) = aT + b\psi$ gives another principal quintic, whose coefficients equal those of the $Q(T,\psi)$-transform. Since the three nonzero transformed coefficients are constrained—that is, each is given by two expressions—they lie on a surface. If this surface were globally parametrized by rational functions of two variables then the $L(T,\psi)$-transformed principal quintic would depend on two parameters, but unfortunately it does not. To proceed, we next establish a general fact about forms, proved by pure linear algebra, leading to a computation that does give a two-parameter quintic.

(5.8.4) PROPOSITION. *Let* $\mathbf{k}$ *be any field not of characteristic 2. In* $\mathbf{k}[X,Y]$, *let* $L = aX + bY$ *be a nonzero linear form and* $Q = \alpha X^2 + 2\beta XY + \gamma Y^2$ *be a quadratic form. Assume* $L \nmid Q$ *in* $\mathbf{k}[X,Y]$. *Let* $m = \alpha b^2 - 2\beta ab + \gamma a^2$ *and* $\delta = \alpha\gamma - \beta^2$. *Then for some linear form* $\tilde{L} = \tilde{a}X + \tilde{b}Y \in \mathbf{k}[X,Y]$,

$$mQ(X,Y) = \tilde{L}(X,Y)^2 + \delta L(X,Y)^2.$$

PROOF. Exercise 5.8.5. □

This result applies to our computations with the principal quintic p. Since $\mathrm{Gal}_{\mathbf{k}}(p) = A_5$ acts transitively on $r_1, \ldots, r_5$, p is irreducible over $\mathbf{k}$ by Proposition 4.3.2. The quotient ring $\mathbf{k}[T]/p\mathbf{k}[T]$ is a field (cf. Section 4.2) and polynomials that are nonzero mod p have polynomial inverses mod p. We are working in full generality, so the form L in Proposition 5.8.3 is nonzero since it is nonzero in specific numerical cases. Thus $L(T,\psi)$ has a polynomial inverse mod p (Exercise 5.8.4), so Proposition 5.8.3 shows that $Q(T,\psi)$ is also invertible mod p and that $L \nmid Q$ in $\mathbf{k}[X,Y]$ (Exercise 5.8.4 again), and the hypothesis of Proposition 5.8.4 is met. Similarly m and δ in Proposition 5.8.3 are nonzero for our L and Q.

To obtain the Brioschi quintic, take L and Q from Proposition 5.8.3 and substitute $X = T$, $Y = \psi$ in Proposition 5.8.4 to get

$$mQ = (\tilde{L} - \sqrt{-\delta}\,L)(\tilde{L} + \sqrt{-\delta}\,L)$$

where now $Q = Q(T,\psi)$, $\tilde{L} = \tilde{L}(T,\psi)$, and $L = L(T,\psi)$ are polynomials in T, so the equality is in $\mathbf{k}(\sqrt{-\delta})[T]$. Proposition 5.8.3 shows that

$$mL \equiv (\tilde{L} - \sqrt{-\delta}\,L)(\tilde{L} + \sqrt{-\delta}\,L) \pmod{p}, \tag{5.8.5}$$

again in $\mathbf{k}(\sqrt{-\delta})[T]$. This is the key congruence for constructing a Tschirnhaus transformation t, given by two different expressions, that is easy to compute with: define

$$t = \tilde{L} \cdot L^{-1} \pmod{p}.$$

Although t lies in the quotient ring $\mathbf{k}[T]/p\mathbf{k}[T]$, we may think of t as any of its representatives in $\mathbf{k}[T]$ since all such representatives act as the same Tschirnhaus transformation of p. To analyze t as a transformation, we continue working mod p. Multiply (5.8.5) by $(mL)^{-1} \pmod{p}$ to obtain the two congruences

$$1 \equiv (t - \sqrt{-\delta})L_1 \pmod{p} \qquad \text{where } L_1 = (\tilde{L} + \sqrt{-\delta}\,L)/m,$$
$$1 \equiv (t + \sqrt{-\delta})L_2 \pmod{p} \qquad \text{where } L_2 = (\tilde{L} - \sqrt{-\delta}\,L)/m.$$

Thus, solving for t gives

$$t \equiv (1+\sqrt{-\delta}\,L_1)L_1^{-1} \pmod p \qquad \text{and} \qquad t \equiv (1-\sqrt{-\delta}\,L_2)L_2^{-1} \pmod p.$$

Since L_1 and L_2 are factors of $Q(T,\psi)$ they are invertible mod p and these expressions for t are defined.

The two expressions for t constrain the t-transformed quintic,

$$q = \prod_{i=1}^{5}(S - t(r_i)) = S^5 + d_1S^4 + d_2S^3 + d_3S^2 + d_4S + d_5.$$

To see this, first note that $q(S) = 0$ for $S = t(r_i) = (1+\sqrt{-\delta}\,L_1(r_i))/L_1(r_i)$, so substituting $S = (1+\sqrt{-\delta}\,R)/R$ expresses q as a rational function of R,

$$q = \frac{(1+\sqrt{-\delta}\,R)^5 + d_1R(1+\sqrt{-\delta}\,R)^4 + d_2R^2(1+\sqrt{-\delta}\,R)^3 + \cdots + d_5R^5}{R^5},$$

which vanishes for $R = L_1(r_i) = (\tilde{L}(r_i,\psi(r_i))+\sqrt{-\delta}\,L(r_i,\psi(r_i))/m$. In other words, the numerator is a quintic in R whose roots are a linear combination of r_i and $\psi(r_i)$; so it is principal by Corollary 5.8.2, and its R^4 and R^3 coefficients vanish,

$$\begin{aligned} 5\delta^2 + 4d_1(-\delta)^{3/2} - 3d_2\delta + 2d_3(-\delta)^{1/2} + d_4 &= 0,\\ 10(-\delta)^{3/2} - 6d_1\delta + 3d_2(-\delta)^{1/2} + d_3 &= 0. \end{aligned}$$

Similarly using the second expression for t, the substitution $S = (1-\sqrt{-\delta}\,R)/R$ expresses q as a rational function of R, again with a principal quintic numerator. But this is the same expression as before except that $\sqrt{-\delta}$ is replaced by $-\sqrt{-\delta}$, so this time the vanishing coefficients are

$$\begin{aligned} 5\delta^2 - 4d_1(-\delta)^{3/2} - 3d_2\delta - 2d_3(-\delta)^{1/2} + d_4 &= 0,\\ -10(-\delta)^{3/2} - 6d_1\delta - 3d_2(-\delta)^{1/2} + d_3 &= 0. \end{aligned}$$

Combining the vanishing conditions gives $5\delta^2 - 3d_2\delta + d_4 = -2d_1\delta + d_3 = -6d_1\delta + d_3 = -10\delta + 3d_2 = 0$. Linear algebra shows $d_1 = d_3 = 0$, $d_2 = \frac{10}{3}\delta$, $d_4 = 5\delta^2$. Letting $C = -\delta/3$,

$$q = S^5 - 10CS^3 + 45C^2S + d_5.$$

This depends on two parameters rather than the three of the original p. To reduce to one parameter, proceed as in Section 4.8: set $\tilde{S} = -SC^2/d_5$ and $W = C^5/d_5^2$; then $q(S) = 0$ if and only if $b(\tilde{S}) = 0$ where b is the Brioschi quintic

$$b = \tilde{S}^5 - 10W\tilde{S}^3 + 45W^2\tilde{S} - W^2.$$

Since the Tschirnhaus transformation t in this section lies in $\mathbf{k}[T]$, it is field-theoretically trivial, i.e., it introduces no irrationalities. Inverting t to convert Brioschi roots to principal roots requires solving a quadratic, cubic, or quartic equation over $\mathbf{k}$.

Though the calculations in this section work in the general quintic environment, they do not specialize to every numerical quintic. For example, Exercise 5.8.1 assumes that $\sigma_3 \neq 0$. Here is a specific example that can be confirmed with the help of a computer algebra system (Exercise 5.8.7). Set $\sigma_1 = \sigma_2 = 0$, $\sigma_3 = -2$, $\sigma_4 = 2$, $\sigma_5 = -1$ to start from the principal quintic

$$p = T^5 + 2T^2 + 2T + 1.$$

Newton's identities show that $s_1 = s_2 = 0$, $s_3 = -6$, $s_4 = -8$, $s_5 = -5$, $s_6 = 12$, $s_7 = 28$, $s_8 = 32$. Thus in Exercise 5.8.1(a), $g = T^3 - (4/3)T^2 + (6/5)$ and $h = T^4 - (5/6)T^2 + (8/5)$. Finding ψ in Exercise 5.8.1(b) requires solving the equation

$$176c_1^2 - 209c_1c_2 - 286c_2^2 = 0,$$

which, nicely, has solution $(c_1, c_2) = (2, 1)$. Thus the result of Proposition 5.8.1 is

$$\psi = T^4 + 2T^3 - (7/2)T^2 + 4.$$

Next, working Exercise 5.8.3 establishes that the result of Proposition 5.8.3 is

$$L(X,Y) = 3960X + 990Y \quad \text{and} \quad Q(X,Y) = -4425X^2 - 120XY + 120Y^2.$$

The values m and δ in Proposition 5.8.4 follow, and $mQ - \delta L^2 = \tilde{L}^2$ with

$$\tilde{L}(X,Y) = 133650(31X + 4Y).$$

With some more arithmetic modulo the quintic p, the Tschirnhaus transformation $t = \tilde{L}(T, \psi(T)) \cdot L(T, \psi(T))^{-1} \pmod{p}$ works out to

$$t = 510T^4 + 120T^3 + 240T^2 + 600T + 960.$$

The t-transformation of p is the resultant $q(S) = R(p(T), S - t(T))$, which is

$$q = S^5 - 1782000S^3 + 1428985800000S - 636613173900000;$$

this is the predicted form of q with $C = 178200$ and $d_5 = -636613173900000$. Finally, taking the resultant $R(q(S), \tilde{S} + SC^2/d_5)$ gives the Brioschi form

$$b = \tilde{S}^5 - (320/72171)\tilde{S}^3 + (5120/578739249)\tilde{S} - (1024/5208653241)$$

with parameter $W = 32/72171$.

Exercises

5.8.1. Prove Proposition 5.8.1 as follows.

(a) Let $g = T^3 - (s_4/s_3)T^2 - (s_3/5)$ and $h = T^4 - (s_5/s_3)T^2 - (s_4/5)$, elements of $\mathbf{k}[T]$. (The s_j are the power sums of the r_i as usual.) Show that

$$\sum g(r_i) = \sum r_i g(r_i) = \sum h(r_i) = \sum r_i h(r_i) = 0.$$

(b) Show that for any $c_1, c_2 \in \mathbf{k}$ the polynomial $\psi = c_1 g + c_2 h$ satisfies

$$\sum \psi(r_i) = \sum r_i \psi(r_i) = 0.$$

Show that choosing c_1 and c_2 to make $\sum \psi(r_i)^2 = 0$ and thus complete the proof requires a square root of

$$-\begin{vmatrix} \sum g(r_i)^2 & \sum g(r_i)h(r_i) \\ \sum g(r_i)h(r_i) & \sum h(r_i)^2 \end{vmatrix}.$$

(c) Since p is principal, Exercise 5.2.5 shows that

$$\Delta(p) = \begin{vmatrix} 5 & 0 & 0 & s_3 & s_4 \\ 0 & 0 & s_3 & s_4 & s_5 \\ 0 & s_3 & s_4 & s_5 & s_6 \\ s_3 & s_4 & s_5 & s_6 & s_7 \\ s_4 & s_5 & s_6 & s_7 & s_8 \end{vmatrix}.$$

Perform row and column operations to reduce this determinant to

$$\begin{vmatrix} 5 & 0 & 0 & 0 & 0 \\ 0 & 0 & s_3 & 0 & 0 \\ 0 & s_3 & s_4 & \sum r_i^2 g(r_i) & \sum r_i^2 h(r_i) \\ 0 & 0 & \sum r_i^2 g(r_i) & \sum g(r_i)^2 & \sum g(r_i)h(r_i) \\ 0 & 0 & \sum r_i^2 h(r_i) & \sum g(r_i)h(r_i) & \sum h(r_i)^2 \end{vmatrix}.$$

This is $5s_3^2$ times the quantity whose square root is required in (b). Thus the square root in (b) lies in $\mathbf{k}$ since $\sqrt{5}$, s_3 and $\sqrt{\Delta(p)}$ do.

5.8.2. Prove Corollary 5.8.2. (Recall that the defining equations of the canonical surface are $\sum Z_i = \sum Z_i^2 = 0$.)

5.8.3. Proving Proposition 5.8.3 requires exhibiting $\alpha, \beta, \gamma, a, b, c \in \mathbf{k}$ such that

$$\alpha T^2 + 2\beta T\psi + \gamma\psi^2 - aT - b\psi + c \equiv 0 \pmod{p} \qquad \text{and } c = 0.$$

(a) Show that the condition $c = 0$ follows from the congruence. (Hint: evaluate the left side of the congruence at r_i and sum over i.) So we only need to establish the congruence.

(b) In Exercise 5.8.1(b), (c_1, c_2) may be taken as $(1,0)$ or $(*,1)$. If $c_2 = 0$, set $\gamma = 1$. Then $\gamma\psi^2$ is quartic mod p, $2\beta T\psi + \gamma\psi^2$ is cubic mod p for appropriate β, $2\beta T\psi + \gamma\psi^2 - b\psi$ is quadratic mod p for appropriate b, etc. If $c_2 = 1$, write $\psi = T^4 + b_1 T^3 + b_2 T^2 + b_4$. Then $T\psi - b_1\psi$ is cubic mod p with leading coefficient $b_2 - b_1^2$. If $b_2 - b_1^2 = 0$, set $2\beta = 1$, $\gamma = 0$, $b = b_1$, and the congruence holds for suitable α, a, c. If $b_2 - b_1^2 \neq 0$ then $\psi^2 - b'\psi$ is cubic mod p for suitable b', so $\psi^2 - b'\psi$ minus a suitable multiple of $T\psi - b_1\psi$ is quadratic mod p, etc. In both cases, specifying $\alpha, \beta, \gamma, a, b, c$, only requires field operations in $\mathbf{k}$.

5.8.4. Assuming $L \neq 0$ in Proposition 5.8.3, show that $L(T, \psi)$ is nonzero (and therefore invertible) mod p, and use the proposition to show that $L \nmid Q$ in $\mathbf{k}[X, Y]$.

5.8.5. Prove Proposition 5.8.4 as follows. Let $M = \begin{bmatrix} \alpha & \beta \\ \beta & \gamma \end{bmatrix}$ and $v = \begin{bmatrix} a \\ b \end{bmatrix}$. these describe Q and L respectively since for any $w \in \mathbf{k}^2$ (viewed as a column vector), $Q(w) = w^t M w$ and $L(w) = v^t w$. In matrix terms the goal of this problem is to show that up to constant multiple, M is congruent to a certain diagonal matrix, by a matrix with bottom row v^t.

Let $\delta = \alpha\gamma - \beta^2$ and $D = \begin{bmatrix} 1 & 0 \\ 0 & \delta \end{bmatrix}$. Introduce the matrix $S = \begin{bmatrix} 0 & 1 \\ -1 & 0 \end{bmatrix}$. Then $S^t = -S$, $S^2 = -I$, and $S^t M S M = \delta I$ so that $M S^t M S M = \delta M$. Set $w = Sv = \begin{bmatrix} b \\ -a \end{bmatrix}$ and $P = \begin{bmatrix} w & SMw \end{bmatrix}$, a 2-by-2 matrix. In the change of variable calculation

$$P^t M P = \begin{bmatrix} w^t \\ w^t M S^t \end{bmatrix} M \begin{bmatrix} w & SMw \end{bmatrix} = \begin{bmatrix} w^t M w & w^t M S M w \\ w^t M S^t M w & w^t M S^t M S M w \end{bmatrix}$$

the off-diagonal entries are equal (since $(P^t M P)^t = P^t M P$, i.e., $P^t M P$ is symmetric) and opposite (since $S^t = -S$, i.e., S is skew), so they are zero. Thus

$$P^t M P = Q(w) \cdot D.$$

From the general formula $\det \begin{bmatrix} x & y \end{bmatrix} = x^t S y$ for vectors x and y, we have $\det P = -Q(w)$, which is nonzero because $L(w) = 0$ and $L \nmid Q$; so P is invertible. Let $m = Q(w)$. (This is the m in the proposition.)

By the formula for the 2-by-2 matrix inverse, $P^{-1} = -\frac{1}{m}U$ where U has bottom row $[a \quad b] = v^t$, and a little more matrix algebra gives

$$mM = U^t DU.$$

This matrix identity proves the proposition.

5.8.6. Confirm the calculation of $d_1, \ldots, d_4$, q, and b.

5.8.7. Confirm the numerical example of the Brioschi reduction at the end of the section.

9. Summary

Figure 5.9.1 displays the field theory of this chapter.

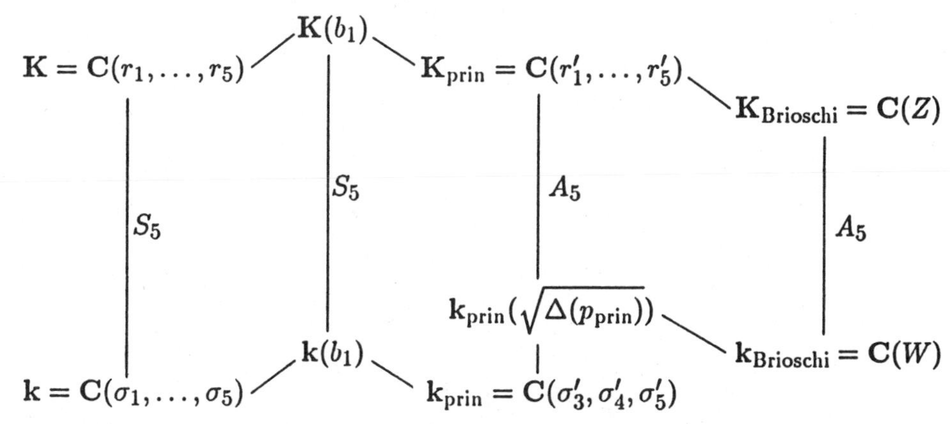

Figure 5.9.1. The Brioschi reduction

The idea is to construct the general quintic extension $\mathbf{K}/\mathbf{k}$, the leftmost extension in the figure. Putting the quintic into depressed form by the Tschirnhaus transformation t_1 in Section 5.4 is field-theoretically trivial over $\mathbf{k}$, so without loss of generality p is depressed. The first step is to adjoin the auxiliary irrationality b_1 from Section 5.4 to the coefficient field $\mathbf{k}$; this immediately takes us out of the splitting field $\mathbf{K}$. Working over $\mathbf{k}(b_1)$, the Tschirnhaus transformation t_2 from Section 5.4 reduces the depressed quintic to principal form p_{prin} with coefficient field $\mathbf{k}_{\text{prin}} = \mathbf{C}(\sigma'_3, \sigma'_4, \sigma'_5)$. Adjoining $\sqrt{\Delta(p_{\text{prin}})}$ reduces the Galois group to A_5. Next, the Tschirnhaus transformation t from Section 5.8 reduces the principal quintic to Brioschi form with coefficient field $\mathbf{k}_{\text{Brioschi}} = \mathbf{C}(W)$.

Once the quintic is reduced all the way down to Brioschi form it is straightforward to solve, assuming an inverse to the icosahedral equation. From Chapter 4, the Brioschi extension is icosahedral, so its splitting field is $\mathbf{K}_{\text{Brioschi}} = \mathbf{C}(Z)$ where $f_I(Z) = W$. Recovering the roots of p from the Brioschi roots amounts to undoing the various Tschirnhaus transformations in reverse order. Discarding the auxiliary quantity finally produces the original splitting field $\mathbf{K}$.

CHAPTER 6

Kronecker's Theorem

The curious feature of reducing the general quintic to one-parameter Brioschi form is the auxiliary b_1. The obvious question is whether the quintic can be reduced to a one-parameter form without using an auxiliary irrationality. Kronecker's Theorem, proved by Klein at the end of [Kl], says this is impossible. The main ideas of the proof (Section 6.2) are clear, but a technical point leads to necessary results from algebra.

Recommended reading: The basics on transcendence degree are in Zariski and Samuel [Za-Sa I]. The proof of Lüroth's Theorem in Section 6.3 follows Jacobson [Ja II]. More general versions of the Embedding Lemma in Section 6.4 appear in papers by Ohm [Oh], Reichstein and Vonessen [Re-Vo], and Roquette [Ro].

1. Transcendence degree

Consider the line $V = \{(z_1, z_2) \in \mathbf{C}^2 : z_1 + z_2 = 0\}$. Although V sits inside the 2-dimensional vector space $\mathbf{C}^2$ it is itself a 1-dimensional vector space over $\mathbf{C}$ in the mathematically precise sense of having basis $\beta = \{(1, -1)\}$ of cardinality 1. Recall that a basis for a vector space V over a field $\mathbf{k}$ is a maximal linearly independent subset of V, i.e., no proper superset is linearly independent; or equivalently (Exercise 6.1.1), a basis is a linearly independent subset that spans V. (For infinite sets, linear independence means that no finite subset satisfies a nonzero polynomial.) In general, the dimension of a vector space V, while defined abstractly as the cardinality of a basis, captures the intuitive notion of how many nonredundant directions V contains.

Blurring definite and indefinite articles by defining "the dimension" of V to be the cardinality of "a basis" is nonsense until we know that all bases have the same cardinality. One way to show this when some basis is finite is the

(6.1.1) DIMENSION LEMMA. *Let $\{x_1, \ldots, x_n\}$ be a basis of the vector space*

V over the field $\mathbf{k}$. *Let $\beta = \{y_1, \dots, y_n, \dots\} \subset V$ be linearly independent over* $\mathbf{k}$ *with $|\beta| \geq n$ (β may be infinite). Then $\{y_1, \dots, y_n\}$ is again a basis for V, so in fact $|\beta| = n$.*

The proof (Exercise 6.1.2) is done by carrying out an exchange process similar to the forthcoming proof of Lemma 6.1.2.

Granting that V has a basis, it is now easy to show that finite dimension is well-defined: if V has a finite basis then the lemma shows that all other bases are finite and that any two finite bases have the same cardinality. (Exercise 6.1.7 asks for the details of this argument in an analogous situation.) Incidentally, existence of a basis for an arbitrary vector space V actually requires a fundamental result called **Zorn's Lemma**, which is equivalent to the Axiom of Choice from set theory—this is an instance where the foundational issues in mathematics meet the undergraduate curriculum.

These counting ideas are general enough to transfer smoothly to field theory. Consider for example the field $\mathbf{K} = \mathbf{C}(x)[y]/\langle y^2 - x^3 - 1\rangle$, where x and y are transcendental over $\mathbf{C}$. (This is a field since $y^2 - x^3 - 1$ is irreducible in $\mathbf{C}(x)[y]$.) While $\mathbf{K}$ is generated over $\mathbf{C}$ by the two elements x and y, these generators are algebraically related, so in some sense $\mathbf{K}$ contains one free variable over $\mathbf{C}$, akin to how the vector space V contained one free direction. Note that $\mathbf{K}$ is algebraic over $\mathbf{C}(x)$ and $\mathbf{C}(y)$, each of which is generated by one transcendental element over $\mathbf{C}$. On the other hand, $\mathbf{K}$ itself does not take the form $\mathbf{C}(z)$ for any z (we will not prove this). In general the appropriate definitions are that a **transcendence base** for an extension $\mathbf{K}/\mathbf{k}$ is a subset of $\mathbf{K}$ that is maximal with respect to algebraic independence over $\mathbf{k}$, or equivalently (Exercise 6.1.3), a transcendence base is an algebraically independent subset β such that $\mathbf{K}/\mathbf{k}(\beta)$ is algebraic; and the **transcendence degree** of $\mathbf{K}/\mathbf{k}$ is the cardinality of a transcendence base.

Analogously to the linear algebra case, Zorn's Lemma guarantees a transcendence base, and finite transcendence degree is well-defined thanks to the

(6.1.2) TRANSCENDENCE DEGREE LEMMA. *Let $\{x_1, \dots, x_n\}$ be a transcendence base for the field extension* $\mathbf{K}/\mathbf{k}$. *Let $\beta = \{y_1, \dots, y_n, \dots\} \subset \mathbf{K}$ be algebraically independent over* $\mathbf{k}$ *with $|\beta| \geq n$ (β may be infinite). Then $\{y_1, \dots, y_n\}$ is again a transcendence base for* $\mathbf{K}/\mathbf{k}$, *so in fact $|\beta| = n$.*

PROOF. Since $y_1 \in \mathbf{K}$ and the extension $\mathbf{K}/\mathbf{k}(x_1, \dots, x_n)$ is algebraic, y_1 occurs in a relation $p_1(y_1, x_1, \dots, x_n) = 0$ for some nonzero polynomial $p_1 \in \mathbf{k}[U_1, T_1, \dots, T_n]$. This relation also must include some x_i, which may be taken to be x_1 after reindexing (Exercise 6.1.4). Thus the extension $\mathbf{k}(y_1, x_1, \ldots, x_n)/\mathbf{k}(y_1, x_2, \ldots, x_n)$ is algebraic; also, $\mathbf{K}/\mathbf{k}(y_1, x_1, \ldots, x_n)$ is algebraic, so since the composite of algebraic extensions is algebraic (showing this is Exercise 6.1.5), exchanging x_1 for y_1 leaves an algebraic extension, $\mathbf{K}/\mathbf{k}(y_1, x_2, \ldots, x_n)$.

Now suppose inductively that for some $k \in \{1, \ldots, n-1\}$, exchanging the variables $x_1, \ldots, x_k$ for $y_1, \ldots, y_k$ leaves an algebraic extension, $\mathbf{K}/\mathbf{k}(y_1, \ldots, y_k, x_{k+1}, \ldots, x_n)$. As in the preceding paragraph, y_{k+1} occurs in a relation $p_{k+1}(y_1, \ldots, y_{k+1}, x_{k+1}, \ldots, x_n) = 0$ which also includes x_{k+1} after reindexing (Exercise 6.1.4). So the extension generated by one more exchange, $\mathbf{K}/\mathbf{k}(y_1, \ldots, y_{k+1}, x_{k+2}, \ldots, x_n)$, is algebraic (Exercise 6.1.4).

By induction, $\mathbf{K}/\mathbf{k}(y_1, \ldots y_n)$ is algebraic. Since $\beta = \{y_1, \ldots, y_n, \ldots\}$ is algebraically independent, the results follow by citing the two definitions of transcendence base. □

It follows (Exercise 6.1.6) that finite transcendence degree is well-defined. (For infinite transcendence degree, which we don't need here, see [Za-Sa I].) The transcendence degree of an extension $\mathbf{K}/\mathbf{k}$ is written $\operatorname{tr}\deg_{\mathbf{k}}(\mathbf{K})$. For example from Chapter 5, $\operatorname{tr}\deg_{\mathbf{C}}(\mathbf{C}(\sigma_1, \ldots, \sigma_5, \sqrt{\Delta(p)})) = 5$. Note that by Exercise 6.1.5, if $\mathbf{k} \subset \mathbf{K} \subset \mathbf{L}$ and $\mathbf{L}/\mathbf{K}$ is algebraic then $\operatorname{tr}\deg_{\mathbf{k}}(\mathbf{L}) = \operatorname{tr}\deg_{\mathbf{k}}(\mathbf{K})$ (Exercise 6.1.7). More generally, transcendence degree is additive: if $\mathbf{k} \subset \mathbf{K} \subset \mathbf{L}$ then (Exercise 6.1.8)

$$\operatorname{tr}\deg_{\mathbf{k}}(\mathbf{L}) = \operatorname{tr}\deg_{\mathbf{K}}(\mathbf{L}) + \operatorname{tr}\deg_{\mathbf{k}}(\mathbf{K}).$$

Exercises

6.1.1. Prove that the text's two definitions of basis for a vector space V over a field $\mathbf{k}$ are equivalent.

6.1.2. Prove the Dimension Lemma.

6.1.3. Prove that the text's two definitions of transcendence base for a field extension $\mathbf{K}/\mathbf{k}$ are equivalent.

6.1.4. Explain why in the proof of the Transcendence Degree Lemma, the relation $p_1(y_1, x_1, \ldots, x_n) = 0$ must involve some x_i. Similarly for the relation $p_{k+1}(y_1, \ldots, y_{k+1}, x_{k+1}, \ldots, x_n) = 0$. Fill in the details as necessary for the induction step of the proof.

6.1.5. Verify that if $\mathbf{L}/\mathbf{K}$ and $\mathbf{K}/\mathbf{k}$ are algebraic extensions then so is $\mathbf{L}/\mathbf{k}$. (Hint: it suffices to show that for every $l \in \mathbf{L}$, the extension $\mathbf{k}(l)/\mathbf{k}$ is finite. But l satisfies some polynomial $p = \sum_{i=0}^{n} a_i T^i \in \mathbf{K}[T]$ and each a_i satisfies a polynomial $p_i \in \mathbf{k}[T] \subset \mathbf{k}(a_0, \ldots, a_{i-1})[T]$, so the chain of finite extensions

$$\mathbf{k} \subset \mathbf{k}(a_0) \subset \cdots \subset \mathbf{k}(a_0, \ldots, a_n) \subset \mathbf{k}(a_0, \ldots, a_n, l)$$

combines with Exercise 4.2.1 to do the trick.)

6.1.6. Prove that finite transcendence degree is well-defined.

6.1.7. Show that if $\mathbf{k} \subset \mathbf{K} \subset \mathbf{L}$ and $\mathbf{L}/\mathbf{K}$ is algebraic then $\operatorname{tr}\deg_{\mathbf{k}}(\mathbf{L}) = \operatorname{tr}\deg_{\mathbf{k}}(\mathbf{K})$.

6.1.8. Prove that transcendence degree is additive. Hint: if $\{x_i\}$ is a transcendence base for $\mathbf{K}$ over $\mathbf{k}$ and $\{y_j\}$ is a transcendence base for $\mathbf{L}$ over $\mathbf{K}$, then $\{x_i\} \cap \{y_j\} = \emptyset$, $\{x_i, y_j\}$ is algebraically independent over $\mathbf{k}$, and the relations

$$\mathbf{L} \subset \overline{\mathbf{K}(\{y_j\})} \subset \overline{\overline{\mathbf{k}(\{x_i\})}(\{y_j\})} = \overline{\mathbf{k}(\{x_i, y_j\})}$$

(the overbars denote algebraic closure) show that $\mathbf{L}/\mathbf{k}(\{x_i, y_j\})$ is algebraic.

2. Kronecker's Theorem

Consider the general set-up of a Tschirnhaus transformation of the quintic. Start as usual with

$$p = \prod_{i=1}^{5}(T - r_i) = \sum_{j=0}^{5}(-1)^j \sigma_j T^{5-j}.$$

Let $\mathbf{k} = \mathbf{C}(\sigma_1, \ldots, \sigma_5, \sqrt{\Delta(p)})$ be the coefficient field of p with the square root of the discriminant adjoined, and let $\mathbf{K} = \mathbf{C}(r_1, \ldots, r_5) = \operatorname{spl}_{\mathbf{k}}(p)$ be the splitting field. Then $\operatorname{Gal}(\mathbf{K}/\mathbf{k}) = A_5$. Take a Tschirnhaus transformation $t \in \mathbf{l}[T]$ where $\mathbf{l}$ is some extension field of $\mathbf{k}$. The corresponding transformed polynomial is

$$q = \prod_{i=1}^{5}(S - t(r_i)).$$

Formulating Kronecker's Theorem precisely requires some care, as the following examples show. First, try the Tschirnhaus transformation $t = T - r_1$. This moves r_1 to zero, so the transformed polynomial is

$$q = S \cdot \prod_{i=2}^{5}(S - t(r_i)) \stackrel{\text{call}}{=} S \cdot \tilde{q}.$$

We can solve the quartic $\tilde{q}$ and the affine t is trivial to invert, so this seems to solve the quintic. Unfortunately, the problem is that the coefficient field of t is $\mathbf{l} = \mathbf{k}(r_1)$, whose construction requires a root of p, so that using this t presupposes a solution to the problem we are trying to solve.

The emerging question is what sort of Tschirnhaus transformations t to consider. Klein and Kronecker, as 19th-century mathematicians, naturally worked over the complex numbers and granted the adjunction of radicals. In the language of field theory, start from any field $\mathbf{f}$, let r be algebraic over $\mathbf{f}$, and suppose $\mathbf{f}(r)/\mathbf{f}$ is constructible by radicals (see Section 4.5). Call such a value r **radical** over $\mathbf{f}$. The **radical closure** of $\mathbf{f}$, written $\mathbf{f}^{\mathrm{rad}}$, is the field generated over $\mathbf{f}$ by all radical values r, a subfield of the algebraic closure of $\mathbf{f}$. With this terminology it is easy to characterize the allowable Tschirnhaus transformations t if we view radical adjunction as trivial: the coefficients of t should be radical over the coefficient field of p, or equivalently, over $\mathbf{k}$. In other words,

$$t \in \mathbf{k}^{\mathrm{rad}}[T].$$

Call such t a **radical** reduction of p.

Now try the radical Tschirnhaus transformation $t = k$ for some $k \in \mathbf{k}$. Since each $t(r_i) = k$, the resulting transformed polynomial is $q = (S - k)^5$. This certainly simplifies the original p, but the catch is that solving p requires inverting the substitution t along with solving q; finding T such that $t(T) = k$ and $p(T) = 0$ is no easier than solving the original equation $p(T) = 0$ and we have accomplished nothing. Note that t is highly degenerate (in the sense of Section 5.5) since it collapses the five roots of p to a common value.

Kronecker's Theorem will consider a radical reduction that introduces only natural irrationalities, meaning $t \in (\mathbf{k}^{\mathrm{rad}} \cap \mathbf{K})[T]$. Since the intersection $\mathbf{k}^{\mathrm{rad}} \cap \mathbf{K}$ is just $\mathbf{k}$ (Exercise 6.2.1), in fact the only relevant Tschirnhaus transformations are field-theoretically trivial,

$$t \in \mathbf{k}[T].$$

Now, if such t collapses the roots of p at all, e.g., $t(r_1) = t(r_2)$, then it quickly follows (Exercise 6.2.2) that all $t(r_i)$ are equal, so t is constant and therefore useless. Thus, to reduce p, we must take t to be nondegenerate. Now we can state

(6.2.1) KRONECKER'S THEOREM. *Let* $\mathbf{K}/\mathbf{k}$ *be the general quintic extension with Galois group* A_5 *and let* $t \in \mathbf{k}[T]$ *be a nondegenerate (i.e., nonconstant) Tschirnhaus transformation taking the general quintic* p *to*

$$q = \prod_{i=1}^{5}(S - \tilde{r}_i) = \sum_{j=0}^{5}(-1)^j \tilde{\sigma}_j S^{5-j}.$$

Then $\operatorname{tr}\deg_{\mathbf{C}}(\mathbf{C}(\tilde{\sigma}_1, \ldots, \tilde{\sigma}_5)) > 1$. *In particular, the coefficients of* q *can not be rational expressions in a single parameter* $w \in \mathbf{k}$.

Certainly $\operatorname{tr}\deg_{\mathbf{C}}(\mathbf{C}(\tilde{\sigma}_1,\ldots,\tilde{\sigma}_5)) > 0$ since t is nonconstant. To prove the theorem, let $\mathbf{F} = \mathbf{C}(\tilde{r}_1,\ldots,\tilde{r}_5) \subset \mathbf{K}$ and $\mathbf{f} = \mathbf{C}(\tilde{\sigma}_1,\ldots,\tilde{\sigma}_5,\sqrt{\Delta(q)}) = \mathbf{F}^{A_5} \subset \mathbf{K}^{A_5} = \mathbf{k}$. (Recall from the Galois Correspondence that the superscript denotes fixed field.) See Figure 6.2.1.

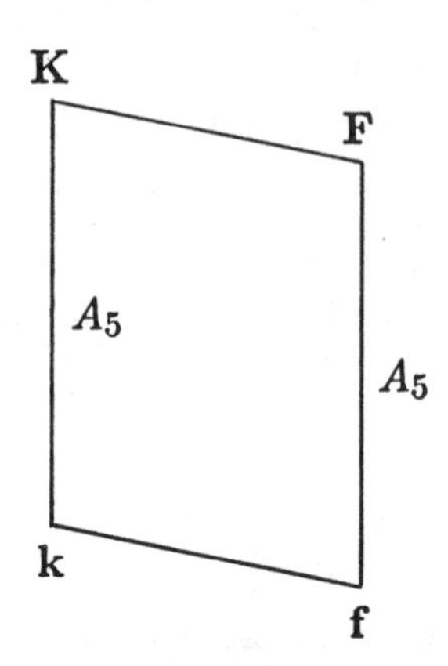

Figure 6.2.1. Fields for the transformed quintic

Since $\mathbf{F}/\mathbf{C}(\tilde{\sigma}_1,\ldots,\tilde{\sigma}_5)$ is algebraic, it suffices to obtain a contradiction by assuming $\operatorname{tr}\deg_{\mathbf{C}}(\mathbf{F}) = 1$. Some algebra, to be developed in the next two sections, shows the crucial fact that this assumption implies $\mathbf{F} = \mathbf{C}(Z)$ for some $Z \in \mathbf{K}$ transcendental over $\mathbf{C}$. Take this as given for now. So $\mathbf{F}$ is generated over $\mathbf{C}$ by a rational expression $Z = \phi_1(r_1,\ldots,r_5)/\phi_2(r_1,\ldots,r_5)$ with $\phi_1,\phi_2 \in \mathbf{k}[T_1,\ldots,T_5]$ and $\gcd(\phi_1,\phi_2) = 1$.

Since $\mathbf{F} = \mathbf{C}(Z)$ and the automorphisms of $\mathbf{C}(Z)$ over $\mathbf{C}$ are $\mathrm{PGL}_2(\mathbf{C})$ (Exercise 4.4.3), the lower Galois group $\mathrm{Gal}(\mathbf{F}/\mathbf{f}) = A_5$ in Figure 6.2.1 gives an injection

$$\rho : A_5 \longrightarrow \mathrm{PGL}_2(\mathbf{C}),$$

i.e., for each $\gamma \in A_5$, $\rho(\gamma) = \overline{\begin{bmatrix} a_\gamma & b_\gamma \\ c_\gamma & d_\gamma \end{bmatrix}} \in \mathrm{PGL}_2(\mathbf{C})$ where $\gamma(Z) = \dfrac{a_\gamma Z + b_\gamma}{c_\gamma Z + d_\gamma}$. (The overbar denotes projective class—the individual matrix entries are only defined up to constant multiple.) Since the automorphisms of $\mathbf{F}$ over $\mathbf{f}$ are restrictions of the automorphisms of $\mathbf{K}$ over $\mathbf{k}$, the map ρ must be compatible with how A_5 acts on $\mathbf{K} = \mathbf{C}(r_1,\ldots,r_5)$ by evenly permuting the r_i. Thus for $\gamma \in A_5$, the quantity $\gamma(Z) = \gamma(\phi_1/\phi_2)$ takes the two forms

(6.2.2)

$$\frac{\phi_1(\gamma(r_1,\ldots,r_5))}{\phi_2(\gamma(r_1,\ldots,r_5))} = \frac{a_\gamma\phi_1(r_1,\ldots,r_5) + b_\gamma\phi_2(r_1,\ldots,r_5)}{c_\gamma\phi_1(r_1,\ldots,r_5) + d_\gamma\phi_2(r_1,\ldots,r_5)}.$$

By Exercise 6.2.3, this is equivalent to the vector equation (with $r_1, \dots, r_5$ now suppressed)

(6.2.3)

$$\begin{bmatrix}\phi_1\\ \phi_2\end{bmatrix} \circ \gamma = k_\gamma \begin{bmatrix}a_\gamma & b_\gamma\\ c_\gamma & d_\gamma\end{bmatrix}\begin{bmatrix}\phi_1\\ \phi_2\end{bmatrix} \qquad \text{for some } k_\gamma \in \mathbf{C}^*.$$

The resulting lift

$$\overline{\begin{bmatrix}a_\gamma & b_\gamma\\ c_\gamma & d_\gamma\end{bmatrix}} \mapsto k_\gamma \begin{bmatrix}a_\gamma & b_\gamma\\ c_\gamma & d_\gamma\end{bmatrix}$$

from $\rho(A_5)$ to $\mathrm{GL}_2(\mathbf{C})$ is well-defined and a homomorphism (Exercise 6.2.4). By Theorems 2.3.1 and 2.6.1, $\rho(A_5)$ is conjugate in $\mathrm{PGL}_2(\mathbf{C})$ to the icosahedral group Γ_I, so conjugating gives a corresponding lift $\Gamma_I \longrightarrow \mathrm{GL}_2(\mathbf{C})$. But no such homomorphism exists (Exercise 6.2.5), giving a contradiction and proving Kronecker's Theorem.

In 1995, Joe Buhler and Zinovy Reichstein proved a much broader theorem ([Bu-Re]) saying, roughly, that the smallest d such that the general polynomial extension $\mathbf{K}/\mathbf{k}$ of degree n can be reduced over $\mathbf{k}$ to transcendence degree d satisfies $\lfloor n/2 \rfloor \le d \le n-3$.

Exercises

6.2.1. Show that $\mathbf{k}^{\mathrm{rad}} \cap \mathbf{K} = \mathbf{k}$ as follows. Each of $\mathbf{k}^{\mathrm{rad}}/\mathbf{k}$ and $\mathbf{K}/\mathbf{k}$ is a normal extension. As cited from [La] in Section 4.3, an algebraic extension $\mathbf{L}/\mathbf{k}$ is normal exactly when any irreducible polynomial $m \in \mathbf{k}[T]$ with a root in $\mathbf{L}$ has all of its roots in $\mathbf{L}$. This criterion shows that the extension $(\mathbf{k}^{\mathrm{rad}} \cap \mathbf{K})/\mathbf{k}$ is normal. Since we are in characteristic zero the extension is therefore Galois. By the Galois Correspondence, $\mathrm{Gal}(\mathbf{K}/(\mathbf{k}^{\mathrm{rad}} \cap \mathbf{K}))$ is a normal subgroup of $\mathrm{Gal}(\mathbf{K}/\mathbf{k}) = A_5$, which is simple (Exercise 4.6.1). Complete the proof.

6.2.2. Taking p, $\mathbf{k}$, $\mathbf{K}$ and $t \in \mathbf{k}[T]$ as in the text, show that if $t(r_1) = t(r_2)$ then all $t(r_i)$, $i = 1, \dots, 5$, are equal. (Hint: apply suitable $\sigma \in A_5$ to the quantity $t(r_1) - t(r_2) = 0_{\mathbf{k}}$.) Show that therefore t is constant. (As discussed in Section 5.4, we only care about $t \pmod p$.)

6.2.3. Prove that (6.2.3) implies (6.2.2). Prove that (6.2.2) implies (6.2.3) as follows.

(a) Given that ϕ_1 and ϕ_2 from the text satisfy $\gcd(\phi_1, \phi_2) = 1$, show that also $\gcd(\phi_1 \circ \gamma, \phi_2 \circ \gamma) = 1$ and $\gcd(a_\gamma\phi_1 + b_\gamma\phi_2, c_\gamma\phi_1 + d_\gamma\phi_2) = 1$ for any $\gamma \in A_5$ and $\left[\begin{smallmatrix}a_\gamma & b_\gamma\\ c_\gamma & d_\gamma\end{smallmatrix}\right] \in \mathrm{GL}_2(\mathbf{C})$.

(b) Complete the proof by applying the lemma from Exercise 5.5.1(b).

6.2.4. Show that the hypothesized map $\rho(A_5) \longrightarrow \mathrm{GL}_2(\mathbf{C})$ in the section is well-defined and must be a homomorphism.

6.2.5. This exercise shows that the representation of Γ_I as a subgroup of $\mathrm{PGL}_2(\mathbf{C})$ does not lift homomorphically to $\mathrm{GL}_2(\mathbf{C})$. Suppose we have such a lifting. Consider the Klein four-group $V = \{1, s_T, t_D, s_T t_D\} \subset \mathrm{PGL}_2(\mathbf{C})$, where as in Section 2.7

$$s_T = \overline{\begin{bmatrix} i & 0 \\ 0 & -i \end{bmatrix}} \quad \text{and} \quad t_D = \overline{\begin{bmatrix} 0 & i \\ i & 0 \end{bmatrix}}.$$

If s_T and t_D lift respectively to $k_s \begin{bmatrix} i & 0 \\ 0 & -i \end{bmatrix}$ and $k_t \begin{bmatrix} 0 & i \\ i & 0 \end{bmatrix}$, show that $s_T t_D$ lifts to $k_s k_t \begin{bmatrix} 0 & -1 \\ 1 & 0 \end{bmatrix}$. Since s_T, t_D and $s_T t_D$ all have order 2, show that necessarily $k_s^2 = k_t^2 = (k_s k_t)^2 = -1$, which is impossible.

Since each Platonic rotation group and each dihedral group D_n for n even contains V or a conjugate copy of V, none of them lifts to $\mathrm{GL}_2(\mathbf{C})$. The cyclic groups and the dihedral groups D_n for n odd do lift.

3. Lüroth's Theorem

To complete the proof of Kronecker's Theorem, we need to show that indeed $\mathbf{F}$ takes the form $\mathbf{C}(Z)$ for some $Z \in \mathbf{K}$, as asserted. Here is the crucial algebraic tool.

(6.3.1) LÜROTH'S THEOREM. *Let* $\mathbf{F}$ *be a field. If* $\mathbf{C} \subset \mathbf{F} \subset \mathbf{C}(r)$ *where* r *is transcendental over* $\mathbf{C}$ *and* $\mathbf{F} \neq \mathbf{C}$, *then* $\mathbf{F}$ *takes the form* $\mathbf{C}(Z)$ *for some* Z *transcendental over* $\mathbf{C}$.

(In fact this theorem and proof hold with $\mathbf{C}$ replaced by an arbitrary field.)

PROOF. There exists some $w \in \mathbf{F} \setminus \mathbf{C}$. It follows (Exercise 6.3.1) that r is algebraic over $\mathbf{C}(w)$ and therefore over $\mathbf{F}$. Let $p_r = \sum_{i=0}^{n} k_i T^{n-i} \in \mathbf{F}[T]$ be the minimal polynomial of r over $\mathbf{F}$. (Thus $k_0 = 1$.)

Each coefficient k_i of p_r lies in $\mathbf{C}(r)$, so (Exercise 6.3.2) there exists a least common denominator $c_0(r) \in \mathbf{C}[r]$ such that $c_0(r)k_i \overset{\text{call}}{=} c_i(r)$ is in $\mathbf{C}[r]$ for $i = 1, \ldots, n$ and $\gcd(c_0(r), \ldots, c_n(r)) = 1$. In other words, $c_0(r)p_r = \sum_{i=0}^{n} c_i(r) T^{n-i} \overset{\text{call}}{=} \tilde{p}$ is primitive in $\mathbf{C}[r][T]$.

Some coefficient k_j of p_r lies outside of $\mathbf{C}$ since r is transcendental over $\mathbf{C}$. So $k_j = g(r)/h(r)$ for polynomials $g, h \in \mathbf{C}[T]$ with $\gcd(g, h) = 1$ and $\max\{\deg(g), \deg(h)\} \stackrel{\text{call}}{=} m \geq 1$. It follows (Exercise 6.3.3) that $[\mathbf{C}(r) : \mathbf{C}(k_j)] = m$; also (Exercise 6.3.3) $\mathbf{C}(k_j) \subset \mathbf{F}$ and $[\mathbf{C}(r) : \mathbf{F}] = n$, so to complete the proof with $Z = k_j$, it suffices to show $m = n$. Note (Exercise 6.3.4) that the r-degree of $\tilde{p} \in \mathbf{C}[r][T]$ is at least m.

Since $k_j = g(r)/h(r)$, r satisfies the polynomial $q_r = g - k_j h \in \mathbf{F}[T]$, so $p_r \mid q_r$ in $\mathbf{F}[T]$, and thus $\tilde{p} \mid q_r$ in the ring $\mathbf{C}(r)[T]$, where $c_0(r)$ is a unit. Substituting $k_j = g(r)/h(r)$ into this divisibility relation and multiplying by $h(r)$ gives

$$g(T)h(r) - g(r)h(T) = h(r)a(r, T)\tilde{p}(r, T) \quad \text{for some } a \in \mathbf{C}(r)[T].$$

Since the left side is in $\mathbf{C}[r][T]$ and $\tilde{p} \in \mathbf{C}[r][T]$ is primitive, this rewrites as

$$g(T)h(r) - g(r)h(T) = b(r, T)\tilde{p}(r, T) \quad \text{for some } b \in \mathbf{C}[r][T].$$

The left side has r-degree at most m and $\tilde{p}(r, T)$ has r-degree at least m, so both r-degrees are exactly m and $b \in \mathbf{C}[T]$. The right side is now primitive in $\mathbf{C}[r][T]$ by Gauss' Lemma (Exercise 4.4.2), hence so is the left side; hence the left side is primitive in $\mathbf{C}[T][r]$ by symmetry, meaning so is the right side. Thus $b \in \mathbf{C}$. Since the left side has equal r-degree and T-degree by symmetry, so does the right side, i.e., $m = n$. □

The hypotheses of Lüroth's Theorem imply that $\operatorname{tr}\deg_{\mathbf{C}}(\mathbf{F}) = 1$ (Exercise 6.3.5). Not all fields of transcendence degree 1 over $\mathbf{C}$ take the form $\mathbf{C}(Z)$, for example $\mathbf{K} = \mathbf{C}(x)[y]/\langle y^2 - x^3 - 1\rangle$ from Section 6.1.

Exercises

6.3.1. In the proof of Lüroth's Theorem, show that r is algebraic over $\mathbf{C}(w)$ and therefore over $\mathbf{F}$.

6.3.2. Show that there exists $c_0(r)$ as claimed in the proof.

6.3.3. Justify the assertion $[\mathbf{C}(r) : \mathbf{C}(k_j)] = m$ in the proof. (See the ideas in Exercise 6.2.3(b) if necessary.) Show that $\mathbf{C}(k_j) \subset \mathbf{F}$ and $[\mathbf{C}(r) : \mathbf{F}] = n$.

6.3.4. Show that the r-degree of $\tilde{p}$ is at least m. (Hint: the degree is at least $\max\{\deg(c_0), \deg(c_j)\}$.)

6.3.5. Show that the hypotheses of Lüroth's Theorem imply $\operatorname{tr}\deg_{\mathbf{C}}(\mathbf{F}) = 1$.

4. The Embedding Lemma

Finally, the following result combines with Lüroth's Theorem to complete the proof of Kronecker's Theorem (Exercise 6.4.1).

(6.4.1) EMBEDDING LEMMA. *Suppose* $\mathbf{F}/\mathbf{C}$ *is a field extension such that* $\operatorname{tr}\deg_{\mathbf{C}}(\mathbf{F}) = 1$, *and there exists a* $\mathbf{C}$*-embedding (meaning an embedding that fixes* $\mathbf{C}$ *pointwise)* $\mathbf{F} \longrightarrow \mathbf{C}(r_1, \ldots, r_{n+1})$ *with* $n \geq 1$ *and* $r_1, \ldots, r_{n+1}$ *algebraically independent over* $\mathbf{C}$. *Then there exists a* $\mathbf{C}$*-embedding* $\mathbf{F} \longrightarrow \mathbf{C}(r_1)$.

To prove this, first make some reductions. Identify $\mathbf{F}$ with its embedded image in $\mathbf{C}(r_1, \ldots, r_{n+1})$. Take a subset $\{s_1, \ldots, s_{n-1}\}$ of $\mathbf{C}(r_1, \ldots, r_{n+1})$ algebraically independent over $\mathbf{F}$ (Exercise 6.4.2 asks why such a set exists). Since transcendence degree is additive, $\operatorname{tr}\deg_{\mathbf{C}}(\mathbf{F}(s_1, \ldots, s_{n-1})) = n = \operatorname{tr}\deg_{\mathbf{C}}(\mathbf{C}(r_1, \ldots, r_n))$. It suffices to give a $\mathbf{C}$-embedding $\mathbf{F}(s_1, \ldots, s_{n-1}) \longrightarrow \mathbf{C}(r_1, \ldots, r_n)$, as restricting then gives a $\mathbf{C}$-embedding $\mathbf{F} \longrightarrow \mathbf{C}(r_1, \ldots, r_n)$ and we are done by induction. After reindexing if necessary, r_{n+1} is transcendental over $\mathbf{F}(s_1, \ldots, s_{n-1})$ (Exercise 6.4.3). Thus, the Embedding Lemma follows from the next result with $\mathbf{L} = \mathbf{F}(s_1, \ldots, s_{n-1})$:

(6.4.2) PROPOSITION. *Suppose* $\mathbf{L}/\mathbf{C}$ *is a field extension with* $\operatorname{tr}\deg_{\mathbf{C}}(\mathbf{L}) = n \geq 1$, *and* $\mathbf{L} \subset \mathbf{C}(r_1, \ldots, r_{n+1})$ *with* r_{n+1} *transcendental over* $\mathbf{L}$. *Then there exists a* $\mathbf{C}$*-embedding* $\mathbf{L} \longrightarrow \mathbf{C}(r_1, \ldots, r_n)$.

PROOF. For any positive integer l the subsets of $\mathbf{L}$

$$R = \{p/q \in \mathbf{L} : p, q \in \mathbf{C}[r_1, \ldots, r_{n+1}], \gcd(p,q) = 1, r_{n+1} - r_1^l \nmid q\}$$

and

$$I = (r_{n+1} - r_1^l)R = \{p/q \in R : r_{n+1} - r_1^l \mid p\}$$

are respectively a ring and a maximal ideal of the ring (Exercise 6.4.4). The substitution $r_{n+1} \mapsto r_1^l$ defines a homomorphism $\varphi : R \longrightarrow \mathbf{C}(r_1, \ldots, r_n)$ whose kernel is I (Exercise 6.4.5). The proof will proceed by showing that $I = \{0\}$ and $R = \mathbf{L}$, so φ is the desired $\mathbf{C}$-embedding. Since $x \in \mathbf{L} \setminus R$ implies $1/x \in I$ (Exercise 6.4.6), showing $I = \{0\}$ also shows that $R = \mathbf{L}$ and therefore suffices for the argument. Thus all we really need to show is that φ is injective.

The image $\varphi(R)$ is a subfield of $\mathbf{C}(r_1, \ldots, r_n)$ (Exercise 6.4.5 again). Call this field $\mathbf{L}^*$. The main step in the proof is to show that $\operatorname{tr}\deg_{\mathbf{C}}(\mathbf{L}^*) = n$. Indeed, note that the extension $\mathbf{C}(r_1, \ldots, r_{n+1})/\mathbf{L}(r_{n+1})$ is algebraic since both fields have transcendence degree $n+1$ over $\mathbf{C}$. So for $i = 1, \ldots, n$ there are polynomial relations

$$a_{i0}r_i^{m_i} + a_{i1}r_i^{m_i-1} \cdots + a_{im_i} = 0 \qquad \text{with each } a_{ij} \in \mathbf{L}[r_{n+1}] \text{ and } a_{i0} \neq 0.$$

(A priori we only know $a_{ij} \in \mathbf{L}(r_{n+1})$, so clear denominators.) Thus each a_{ij} takes the form

$$a_{ij} = b_{ij0}r_{n+1}^{m_{ij}} + b_{ij1}r_{n+1}^{m_{ij}-1} + \cdots + b_{ijm_{ij}} \qquad \text{with each } b_{ijk} \in \mathbf{L} \text{ and } b_{ij0} \neq 0.$$

Consider the coefficient set $C = \{a_{ij}, b_{ijk}\} \subset \mathbf{C}(r_1, \ldots, r_{n+1})$. For large enough l (in particular, we may assume $l > m_1$), the polynomial $r_{n+1} - r_1^l$ divides no numerator or denominator of any nonzero element of C. Applying φ to C (i.e., substituting $r_{n+1} \mapsto r_1^l$) gives a new set $C' = \{a'_{ij}, b'_{ijk}\} \subset \mathbf{C}(r_1, \ldots, r_n)$ and new polynomial relations for $i = 1, \ldots, n$,

$$a'_{i0}r_i^{m_i} + a'_{i1}r_i^{m_i-1} \cdots + a'_{im_i} = 0 \qquad \text{with each } a'_{ij} \in \mathbf{L}^*[r_1^l] \text{ and } a'_{i0} \neq 0.$$

Thus $\mathbf{C}(r_1, \ldots, r_n)/\mathbf{L}^*(r_1)$ is algebraic. Also, the relation

$$\begin{aligned} 0 &= a'_{10}r_1^{m_1} + \cdots + a'_{1m_1} \\ &= (b'_{100}r_1^{lm_{10}} + \cdots + b'_{10m_{10}})r_1^{m_1} + \cdots + (b'_{1m_10}r_1^{lm_{1m_1}} + \cdots) \end{aligned}$$

is nontrivial (Exercise 6.4.7), so $\mathbf{L}^*(r_1)/\mathbf{L}^*$ is algebraic too. Concatenating extensions shows that $\mathbf{C}(r_1, \ldots, r_n)/\mathbf{L}^*$ is algebraic, so that $\operatorname{tr deg}_{\mathbf{C}}(\mathbf{L}^*) = n$ as claimed.

Now we can prove that φ is injective. Let $\{y_1, \ldots, y_n\}$ be a transcendence base for $\mathbf{L}^*/\mathbf{C}$ and let $\{x_1, \ldots, x_n\} \subset R$ satisfy $\varphi(x_i) = y_i$ for each i. Then $\{x_1, \ldots, x_n\}$ is algebraically independent over $\mathbf{C}$ (Exercise 6.4.8) and therefore a transcendence base for $\mathbf{L}/\mathbf{C}$. Each nonzero $x \in R$ satisfies a polynomial relation

(6.4.3)

$$\sum_{i=0}^{d} p_i(x_1, \ldots, x_n)x^i = 0 \qquad \text{with each } p_i \in \mathbf{C}[T_1, \ldots, T_n] \text{ and } p_0 \text{ nonzero}$$

(Exercise 6.4.9). Applying φ to both sides gives $\sum_{i=0}^{d} p_i(y_1, \ldots, y_n)(\varphi x)^i = 0$. Since $p_0(y_1, \ldots, y_n) \neq 0$ (Exercise 6.4.10), $\varphi x \neq 0$. Thus φ is injective and the proof is complete. $\square$

Exercises

6.4.1. Use the Embedding Lemma and Lüroth's Theorem to complete the proof of Kronecker's Theorem.

6.4.2. Under the hypotheses of the Embedding Lemma, explain why there is a set $\{s_1, \ldots, s_{n-1}\}$ as described.

6.4.3. Under the hypotheses of the Embedding Lemma, show that some r_i (which may as well be r_{n+1}) is transcendental over $\mathbf{F}(s_1, \ldots, s_{n-1})$.

6.4.4. Show that the sets R and I in Proposition 6.4.2 are a ring and a maximal ideal of the ring. (For the second part, show every element of $R \setminus I$ is a unit in R.)

6.4.5. Show that the substitution $r_{n+1} \mapsto r_1^l$ defines a homomorphism $\varphi : R \longrightarrow \mathbf{C}(r_1, \ldots, r_n)$ whose kernel is I. (In fancy language, R and I are the **valuation ring** and **maximal ideal** of the $(r_{n+1} - r_1^l)$**-adic valuation on L**. This valuation, $\nu : R \longrightarrow \mathbf{N} \cup \infty$, returns the exponent of $r_{n+1} - r_1^l$ in each element of R.) Show that $\varphi(R)$ is a subfield of $\mathbf{C}(r_1, \ldots, r_n)$.

6.4.6. Show that if $x \in \mathbf{L} \setminus R$ then $1/x \in I$, so in particular showing that $I = \{0\}$ also shows that $R = \mathbf{L}$.

6.4.7. Show that the relation $0 = (b'_{100}r_1^{lm_{10}} + \cdots + b'_{10m_{10}})r_1^{m_1} + \cdots$ is nontrivial by considering the coefficient of $r_1^{lm_{10}+m_1}$. (Recall that $l > m_1$.)

6.4.8. Show that $\{x_1, \ldots, x_n\}$ is algebraically independent over $\mathbf{C}$.

6.4.9. Show that each nonzero $x \in R$ satisfies a polynomial relation (6.4.3).

6.4.10. Show that $p_0(y_1, \ldots, y_n) \neq 0$.

5. Summary

The proof of Kronecker's Theorem has two components: the easy part is that the subgroup Γ_I of $\mathrm{PGL}_2(\mathbf{C})$ isn't the projective image of an isomorphic subgroup of $\mathrm{GL}_2(\mathbf{C})$; the hard part is the algebra in Lüroth's Theorem and its extension to the Embedding Lemma. The next chapter will present a broad view of the connections between algebraic equations and geometry, which will illuminate the geometric ideas underlying the harder algebra here. See in particular Exercise 7.2.10.

CHAPTER 7

Computable extensions

Granting the adjunction of radicals, Chapter 5 showed that solving the general quintic is equivalent to solving the Brioschi quintic

$$b_{W'} = T^5 - 10W'T^3 + 45W'^2T - W'^2,$$

a process that Section 4.8 carried out with the solution to the icosahedral equation $f_I(Z) = W$, where W' and W are related by the fractional linear transformation $W' = 1/(1728(1 - W))$. This chapter solves the Brioschi quintic by a different method: repeatedly iterating a rational function. Not surprisingly, finding the rational function to iterate relies on icosahedral geometry.

Recall that the Radical Criterion from Chapter 4 gives a group-theoretic characterization of field extensions that can be constructed by successively adjoining radicals. Analogously, this chapter describes extensions that can be constructed by successively iterating rational functions. Finding the iteration for the Brioschi quintic will do the hard work of establishing the characterization.

Recommended reading: To keep the book manageable in scope, this chapter quotes results and skips technicalities from algebraic geometry and complex dynamics. The material is drawn closely from the paper of Doyle and McMullen [Do-Mc]; see the paper and its references for the theorems from dynamics cited here without proof. The compressed development of algebraic geometry in Section 7.2 is meant to make the material accessible without going into full detail, so supplementing with a text such as Reid [Reid], Cox, Little, and O'Shea [Co-Li-O'S], Mumford [Mu], or the first few sections of Hartshorne [Ha] may be helpful.

1. Newton's method for nth roots

For any polynomial $p \in \mathbf{C}[T]$, **Newton's method** for finding roots of p is: Make an initial guess t_0, and then generate a sequence of t-values by the iteration

$$t_{k+1} = t_k - \frac{p(t_k)}{p'(t_k)} \qquad \text{for } k = 0, 1, 2, \ldots$$

(See Exercise 7.1.1 for the geometric idea behind this formula.) That is, apply the function $T - p/p' \in \mathbf{C}(T)$ repeatedly, starting at t_0. The hope is that the resulting sequence $\{t_k\}$ will converge to a root of p. If the initial guess t_0 is close enough to a simple (nonrepeating) root, this will indeed happen. Newton's method is **decision-free**, meaning that the iteration can be coded as a computer algorithm without any if-then statements. Of course, making Newton's method a true algorithm by stopping the iteration after finitely many steps requires a decision, but that is a separate issue. Think of generating the sequence $\{t_k\}$ as a single decision-free process.

In particular, to find nth roots of unity by iteration, apply Newton's method to the polynomial $p = T^n - 1$. For this special case, the function being iterated to produce the sequence $\{t_k\}$ is (Exercise 7.1.2)

$$f = \frac{(n-1)T^n + 1}{nT^{n-1}} \in \mathbf{C}(T).$$

A fact from dynamics, to be shown later in Exercise 7.8.8, is that Newton's method for nth roots of unity is **generally convergent**. This means that for any t_0 from a dense open set of initial guesses $D \subset \widehat{\mathbf{C}}$ (meaning $\overline{D} = \widehat{\mathbf{C}}$, where the overbar denotes smallest closed superset), iterating f on t_0 produces a sequence $\{t_k\}$ that converges to a root of p. For most polynomials p, Newton's method is not generally convergent.

Iterating f to find the nth roots of unity is one of a family of algorithms for finding the nth roots of nonzero complex numbers w, i.e., finding the roots of the polynomials $p_w = T^n - w$. For each w, Newton's method is to iterate the function

$$F_w = \frac{(n-1)T^n + w}{nT^{n-1}} \in \mathbf{C}(T)$$

(Exercise 7.1.2 again). Viewing all the nth root problems collectively as a w-parametrized family leads to the all-encompassing two-variable rational function

$$F = F_W = \frac{(n-1)T^n + W}{nT^{n-1}} \in \mathbf{C}(W)(T)$$

where as usual the upper-case W is a formal symbol. The idea is to think of F as a one-parameter family $\{F_w\}$ of rational functions in the iteration variable T. Equivalently, rather than considering F as an element of $\mathbf{C}(W)(T)$, one may view it as a function $F : \widehat{\mathbf{C}} \setminus \{0, \infty\} \longrightarrow \mathbf{C}(T)$ taking each w to the appropriate rational function F_w.

The configuration of nth roots of any nonzero w is conformally equivalent to the configuration of nth roots of unity from the special case $w = 1$, when $p_w = p_1$ is the p from above; specifically, for any z such that $z^n = w$, a fractional linear transformation taking the nth roots of w to the nth roots of unity is simply the scaling

$$\phi_z = \begin{bmatrix} 1/z & 0 \\ 0 & 1 \end{bmatrix} : t \mapsto \frac{t}{z}$$

(Exercise 7.1.3). This suggests that Newton's method for any p_w (i.e., iterating F_w) is conformally equivalent to Newton's method for p (i.e., iterating f). And indeed, if the symbols Z and W are related by $Z^n = W$, define

$$\phi = \phi_Z = \begin{bmatrix} 1/Z & 0 \\ 0 & 1 \end{bmatrix} \in \mathrm{PGL}_2(\mathbf{C}(Z)),$$

representing the Z-dependent rational function $T/Z \in \mathbf{C}(Z)(T)$, with inverse $\phi^{-1} = \left[\begin{smallmatrix} Z & 0 \\ 0 & 1 \end{smallmatrix}\right] = ZT$; then

$$\begin{aligned}(\phi_Z^{-1} \circ f \circ \phi_Z)(T) = Z \cdot f(T/Z) = Z \cdot \frac{(n-1)(T/Z)^n + 1}{n(T/Z)^{n-1}} &= \frac{(n-1)T^n + W}{nT^{n-1}} \\ &= F_W(T).\end{aligned}$$

Thus $F = \phi^{-1} f \phi$. (The composition symbol "$\circ$" will generally be omitted from now on for brevity.) This equation says that Newton's method for the polynomials p_w is a **rigid family** of iterations: the **model** $f \in \mathbf{C}(T)$ and the **conjugating conformal map** $\phi \in \mathrm{PGL}_2(\mathbf{C}(Z))$ give the general iteration $F \in \mathbf{C}(W)(T)$. Since the model f is generally convergent, so is any iteration F_w (Exercise 7.1.4).

(A small point: the conjugating transformation ϕ does not lie in the group $\mathrm{PSL}_2(\mathbf{C}(Z))$; unlike $\mathrm{PGL}_2(\mathbf{C})$ and $\mathrm{PSL}_2(\mathbf{C})$, the groups $\mathrm{PGL}_2(\mathbf{C}(Z))$ and $\mathrm{PSL}_2(\mathbf{C}(Z))$ are not isomorphic since the field $\mathbf{C}(Z)$ is not closed under square roots—cf. Exercise 2.1.4. For uniform notation we will also write $\mathrm{PGL}_2(\mathbf{C})$ rather than $\mathrm{PSL}_2(\mathbf{C})$ for the next few sections.)

An output value of Newton's method F is a point $(w, z) \in \widehat{\mathbf{C}} \times \widehat{\mathbf{C}}$ with $z^n = w$. One may picture the set of such outputs as the graph of the n-valued function $\sqrt[n]{}$ on the w-sphere, sitting inside $\widehat{\mathbf{C}} \times \widehat{\mathbf{C}}$. Newton's method F receives a point $(w, t) \in \widehat{\mathbf{C}} \times \widehat{\mathbf{C}}$, where we want an nth root of w and t is an initial guess; iterating F_w will reliably move the point (w, t) vertically to a point (w, z) on the graph. See Figure 7.1.1.

Computationally, the field $\mathbf{C}(W)$ is the set of algorithms that take input from the w-sphere and perform finitely many rational operations. That

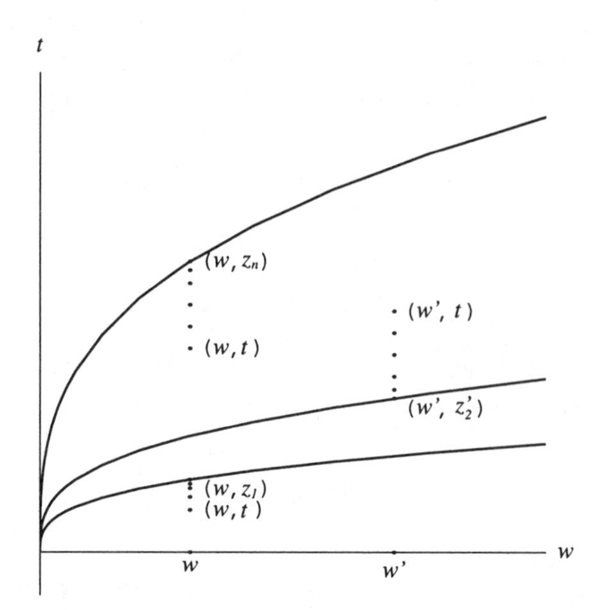

Figure 7.1.1. Newton's method for nth roots of various w

is, evaluating any particular rational function requires finitely many adds, subtracts, multiplies, and divides; and only a rational function can be so evaluated. Similarly, the extension field $\mathbf{C}(Z)$ is the set of algorithms that perform finitely many rational operations on input from the z-sphere, which maps to the w-sphere by $z \mapsto z^n$. But—and this leads to a central idea of this chapter—the field $\mathbf{C}(Z)$ can also be described another way as a set of algorithms: it consists of the algorithms that first take input from the w-sphere, then call Newton's method for nth roots to get a value z such that $z^n = w$, and then perform finitely many rational operations on z. That is, "adjoining" Newton's method for nth roots to the operations of $\mathbf{C}(W)$ gives the operations of $\mathbf{C}(Z)$ and thus constructs the field extension $\mathbf{C}(Z)/\mathbf{C}(W)$ computationally. This new idea is crucial in the sequel: rather than constructing field extensions algebraically by adjoining roots, we will now construct them computationally by adjoining algorithms (in this case, Newton's method for nth roots) to the rational operations of the base field.

The rigidity of F—i.e., the equality $F = \phi^{-1} f \phi$—imposes conditions on the conjugating transformation ϕ. To wit, $\phi = T/Z$ is a rational function of T with coefficient $1/Z$, hence the composition $F = \phi^{-1} f \phi$ lies in $\mathbf{C}(Z)(T)$; but from before we know the stronger fact that F actually lies in the subfield $\mathbf{C}(W)(T)$. Thus, F remains invariant when its coefficients are transformed under the Galois group $\Gamma = \mathrm{Gal}(\mathbf{C}(Z)/\mathbf{C}(W))$, which is the cyclic group of order n with generator $Z \mapsto \zeta_n Z$. In other words, $(\phi^{-1} f \phi)^\gamma = \phi^{-1} f \phi$ for all

$\gamma \in \Gamma$, or, since γ is a homomorphism and the coefficients of f are constant,

$$(\phi^{-1})^\gamma f \phi^\gamma = \phi^{-1} f \phi \quad \text{for all } \gamma \in \Gamma. \tag{7.1.1}$$

To study this condition further, define

$$\rho(\gamma) = \phi^\gamma \phi^{-1} \in \mathrm{PGL}_2(\mathbf{C}(Z))$$

for each $\gamma \in \Gamma$. We now investigate $\rho(\gamma)$, using fairly general methods. Of course, it is simple to compute directly that for any $\gamma : Z \mapsto \zeta_n^j Z$,

$$\rho(\gamma) = \phi_Z^\gamma \phi_Z^{-1} = \begin{bmatrix} 1/(\zeta_n^j Z) & 0 \\ 0 & 1 \end{bmatrix} \begin{bmatrix} Z & 0 \\ 0 & 1 \end{bmatrix} = \begin{bmatrix} \zeta_n^{-j} & 0 \\ 0 & 1 \end{bmatrix} : T \mapsto T/\zeta_n^j;$$

but the general methods will carry over nicely to a broader context in Section 7.3, so they are being demonstrated here specifically for Newton's method to make the later reading easier.

First, by Exercise 7.1.5, the Γ-invariance (7.1.1) of F is equivalent to the condition

$$\rho(\gamma)^{-1} f \rho(\gamma) = f \quad \text{for all } \gamma \in \Gamma; \tag{7.1.2}$$

that is, each $\rho(\gamma)$ commutes with f. This is easy to confirm explicitly from the computation of $\rho(\gamma)$ above (Exercise 7.1.6).

Next we rederive that $\rho(\gamma)$ in fact lies in $\mathrm{PGL}_2(\mathbf{C})$ rather than in the larger $\mathrm{PGL}_2(\mathbf{C}(Z))$. For any nonzero $z, w \in \mathbf{C}$ with $z^n = w$, we know that the transformation $\phi_z \in \mathrm{PGL}_2(\mathbf{C})$ maps the nth roots of w to the set of nth roots of unity,

$$A = \{1, \zeta_n, \zeta_n^2, \ldots, \zeta_n^{n-1}\}.$$

Similarly, for any $\gamma : Z \mapsto \zeta_n^j Z$ in Γ, the transformation

$$\phi_z^\gamma = \begin{bmatrix} 1/(\zeta_n^j z) & 0 \\ 0 & 1 \end{bmatrix} : t \mapsto \frac{t}{\zeta_n^j z}$$

maps the nth roots of w to A. It follows that the transformation $\rho(\gamma)_z = \phi_z^\gamma \phi_z^{-1}$ permutes A. This also follows directly from (7.1.2) since A consists of the fixed points of f in $\mathbf{C}$ (Exercise 7.1.7). In any case, since A is a discrete subset of $\widehat{\mathbf{C}}$, the transformation $\rho(\gamma)_z$ permutes A in the same way as z varies continuously. Any two z-values in $\widehat{\mathbf{C}} \setminus \{0, \infty\}$ are joined by a path, so the general fractional linear transformation $\rho(\gamma)$ permutes A independently of Z. Since fractional linear transformations act triply transitively on the sphere (see for example Section 2.5 of [Jo-Si]), $\rho(\gamma)$ is determined by its action on A and therefore lies in $\mathrm{PGL}_2(\mathbf{C})$ as claimed. Note that this proof tacitly assumes $n \geq 3$ and omits the square root case.

Now compute for any $\gamma_1, \gamma_2 \in \Gamma$,

$$\begin{aligned}
\rho(\gamma_1\gamma_2) &= \phi^{\gamma_1\gamma_2}\phi^{-1} = \phi^{\gamma_1\gamma_2}(\phi^{\gamma_2})^{-1}\phi^{\gamma_2}\phi^{-1} \\
&= \phi^{\gamma_1\gamma_2}(\phi^{-1})^{\gamma_2}\rho(\gamma_2) && \text{by the hint to Exercise 7.1.5} \\
&= (\phi^{\gamma_1}\phi^{-1})^{\gamma_2}\rho(\gamma_2) && \text{because } \gamma_2 \text{ is a homomorphism} \\
&= \rho(\gamma_1)^{\gamma_2}\rho(\gamma_2) = \rho(\gamma_1)\rho(\gamma_2) && \text{because } \rho(\gamma_1) \text{ is } Z\text{-independent.}
\end{aligned}$$

This shows that $\rho : \Gamma \longrightarrow \mathrm{PGL}_2(\mathbf{C})$ is a homomorphism. Finally, ρ is injective because for any $\gamma \in \Gamma$ taking $Z \mapsto \zeta_n^j Z$ and any nonzero $z \in \mathbf{C}$,

$$\phi_z(z) = \begin{bmatrix} 1/z & 0 \\ 0 & 1 \end{bmatrix}(z) = 1 \quad \text{and} \quad \phi_z^\gamma(z) = \begin{bmatrix} 1/(\zeta_n^j z) & 0 \\ 0 & 1 \end{bmatrix}(z) = \zeta_n^{-j};$$

thus $\rho(\gamma) = \phi_z^\gamma \phi_z^{-1}$ takes $1 \mapsto z \mapsto \zeta_n^{-j}$, showing that $\rho(\gamma)$ permutes A nontrivially unless γ is the identity mapping. Of course, this is all clear from the explicit value of ρ (Exercise 7.1.6 again).

To summarize, Newton's method $F \in \mathbf{C}(W)(T)$ for the family of polynomials $p_w = T^n - w$ is generally convergent and therefore constructs the field extension $\mathbf{C}(Z)/\mathbf{C}(W)$ computationally, where $Z^n = W$. It is rigid, meaning it has the form $F = \phi^{-1}f\phi$ with $f \in \mathbf{C}(T)$ and $\phi \in \mathrm{PGL}_2(\mathbf{C}(Z))$. Consequently the map $\gamma \mapsto \phi^\gamma\phi^{-1}$ for each $\gamma \in \Gamma = \mathrm{Gal}(\mathbf{C}(Z)/\mathbf{C}(W))$ is an embedding $\rho : \Gamma \longrightarrow \mathrm{Aut}(f)$, where

$$\mathrm{Aut}(f) = \{m \in \mathrm{PGL}_2(\mathbf{C}) : m^{-1}fm = f\}.$$

Exercises

7.1.1. Show that the formula for Newton's method comes from intersecting the tangent to the graph of p at $(t_k, p(t_k))$ with the t-axis for the next approximation t_{k+1}. (See Figure 7.1.2.)

7.1.2. Verify that Newton's method for $p = T^n - 1$ is to iterate the function $f = ((n-1)T^n + 1)/(nT^{n-1})$, and more generally that Newton's method for $p_w = T^n - w$ is to iterate the function $F_w = ((n-1)T^n + w)/(nT^{n-1})$.

7.1.3. Confirm that the fractional linear transformation ϕ_z takes the nth roots of w to the nth roots of unity.

7.1.4. Use the fact that f is generally convergent and the relation $F = \phi^{-1}f\phi$ to show that any iteration F_w for $w \in \widehat{\mathbf{C}} \setminus \{0, \infty\}$ is also generally convergent. (The iteration $\{t, f(t), f(f(t)), \ldots\} = \{f^k(t)\}$ converges for all initial guesses t in a dense open set $D \subset \widehat{\mathbf{C}}$. Show that the iteration $\{F_w^k(t)\}$ converges for all $t \in \phi_z^{-1}(D)$ for any z such that

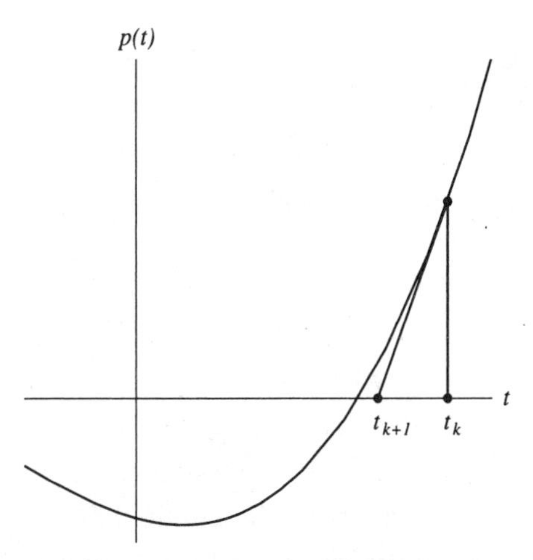

Figure 7.1.2. Geometry of Newton's method

$z^n = w$. Note that $\phi(\overline{\phi^{-1}(D)})$ is a closed superset of $\phi(\phi^{-1}(D)) = D$ and therefore a superset of $\overline{D} = \widehat{\mathbf{C}}$. Apply ϕ^{-1}.)

7.1.5. Show that the Γ-invariance of F is equivalent to the conjugacy-invariance condition $\rho(\gamma)^{-1} f \rho(\gamma) = f$ for all $\gamma \in \Gamma$. (Observe first that since γ is a homomorphism, $(\phi^{-1})^\gamma \phi^\gamma = (\phi^{-1}\phi)^\gamma = 1^\gamma = 1$, showing that $(\phi^\gamma)^{-1} = (\phi^{-1})^\gamma$.)

7.1.6. Use the computed value of $\rho(\gamma)$ to verify explicitly that it commutes with f and that $\rho : \Gamma \longrightarrow \mathrm{PGL}_2(\mathbf{C})$ is an embedding.

7.1.7. (a) Show that the relation $\rho(\gamma)^{-1} f \rho(\gamma) = f$ implies that $\rho(\gamma)$ permutes the fixed points of f.

(b) If $p \in \mathbf{C}[T]$ has only simple roots, show that the fixed points of the Newton iteration $f = T - p/p'$ in $\mathbf{C}$ are the roots of p. In particular, the fixed points of f from the text are the nth roots of unity.

2. Varieties and function fields

Analogously to Newton's method for nth roots, we seek a purely iterative algorithm $F \in \mathbf{C}(W')(T)$ such that iterating each $F_{w'} \in \mathbf{C}(T)$ will generate a root of the Brioschi quintic $b_{w'}$. To investigate whether such an algorithm exists, we need to characterize the field extensions that are constructible by decision-free iteration, just as Galois theory characterizes the field extensions that are constructible by radical adjunction. The first step toward discussing this problem in its proper environment is to expand our vocabulary to generalize the connection between the z-sphere, the w-sphere, and their fields of rational functions $\mathbf{C}(Z)$ and $\mathbf{C}(W)$.

As the preceding section mentioned, the function $z \mapsto z^n$ maps the z-sphere to the w-sphere, and this map corresponds to the field extension $\mathbf{C}(Z)/\mathbf{C}(W)$ where $Z^n = W$. The cyclic group C_n describes both the w-invariant automorphisms of the z-sphere and the Galois group of the corresponding field extension. Similarly for any finite rotation group Γ from Chapter 2, the Γ-invariant function f_Γ from Chapter 3 maps the z-sphere to the w-sphere, and as shown in Chapter 4, Γ is also the Galois group of the field extension $\mathbf{C}(Z)/\mathbf{C}(W)$, where now $f_\Gamma(Z) = W$. In all of these cases, the group Γ acts geometrically (from the left) as the motions of the z-sphere that fix the f_Γ-image w-sphere, and it acts algebraically (from the right) as the automorphisms of $\mathbf{C}(Z)$ that fix the subfield $\mathbf{C}(W)$.

Unfortunately, these examples aren't broad enough to serve as the basis for a good theory. We need to discuss mappings between sets other than spheres and extensions of fields other than the rational functions of one variable. Here are two more scenarios where a group describes both the geometry of a set-mapping and the algebra of an associated field extension.

Let $p \in \mathbf{C}[T]$ be cubic with distinct roots and consider the **elliptic curve**

$$\mathcal{E} = \{(w, z) \in \mathbf{C}^2 : z^2 = p(w)\}.$$

The corresponding rational functions taking input from $\mathcal{E}$ are

$$K(\mathcal{E}) = \mathbf{C}(W)[Z]/\langle Z^2 - p(W)\rangle$$

where $\langle Z^2 - p(W)\rangle$ is the ideal of the polynomial ring $\mathbf{C}(W)[Z]$ generated by all multiples of $Z^2 - p(W)$. This ideal is maximal since $Z^2 - p(W)$ is irreducible over $\mathbf{C}(W)$, so the quotient is indeed a field. The elliptic curve maps to the w-sphere by $(w, z) \mapsto w$, and the corresponding field extension is $K(\mathcal{E})/\mathbf{C}(W)$. The cyclic group C_2 describes the w-invariant motions of the elliptic curve (they are the identity and $(w, z) \mapsto (w, -z)$) and the Galois group of the field extension. Thus once again we have a map between sets and a corresponding extension of fields, both described by the same group C_2; but this example differs from the preceding ones in that the elliptic curve $\mathcal{E}$ is not a sphere—in fact, it turns out to be a topological torus with one point missing—and (as alluded in Chapter 6) its field of rational functions $K(\mathcal{E})$ does not take the form $\mathbf{C}(Z)$ for any symbol Z.

The function field of the elliptic curve is still fairly specialized since it stays in the realm of transcendence degree 1. For an example of higher transcendence degree, the map $\mathbf{C}^n \longrightarrow \mathbf{C}^n$ from general polynomial roots to coefficients, $(r_1, \ldots, r_n) \mapsto (\sigma_1, \ldots, \sigma_n)$, corresponds to the general polynomial extension field from Chapter 5; and the group S_n describes all motions of the root space that fix the coefficient space as well as being the Galois group of the field extension.

This section introduces some terminology from **algebraic geometry** to describe a broad class of sets, fields, and groups that interrelate in the fashion of these examples.

Let n be a positive integer.

Each ideal I in the polynomial ring $\mathbf{C}[Z_1,\dots,Z_n]$ has an associated **algebraic set** in $\mathbf{C}^n$,

$$V(I) = \{(z_1,\dots,z_n) \in \mathbf{C}^n : g(z_1,\dots,z_n) = 0 \text{ for all } g \in I\}.$$

In words, the algebraic set of an ideal is the points where every polynomial in the ideal vanishes. The empty set $\emptyset$ and all of $\mathbf{C}^n$ are algebraic. If $V(I_1)$ and $V(I_2)$ are algebraic sets then their union is $V(I_1I_2)$ (where I_1I_2 is the ideal generated by products g_1g_2 with $g_1 \in I_1$ and $g_2 \in I_2$), another algebraic set. If $\{V(I_j) : j \in J\}$ is an arbitrary collection of algebraic sets, their intersection is $V(\langle I_j : j \in J\rangle)$ (where $\langle I_j : j \in J\rangle$ is the ideal generated by all the I_j), again an algebraic set. (See Exercise 7.2.1 for all this.) Thus the algebraic sets may be viewed as the closed sets in the **Zariski topology** on $\mathbf{C}^n$, and each algebraic set becomes a topological space in its own right under the induced topology.

Each set S in $\mathbf{C}^n$ has in turn an **associated ideal** in $\mathbf{C}[Z_1,\dots,Z_n]$,

$$I(S) = \{g \in \mathbf{C}[Z_1,\dots,Z_n] : g(z_1,\dots,z_n) = 0 \text{ for all } (z_1,\dots,z_n) \in S\}.$$

In words, the ideal of a set is the polynomials that vanish at every point in the set. For any set S, $V(I(S))$ is the smallest algebraic superset of S (i.e., its Zariski closure), denoted $\overline{S}$, and $I(V(I(S))) = I(S)$, i.e., $I(\overline{S}) = I(S)$. If $S_1 \subset S_2$ is a proper containment of algebraic sets then $I(S_1) \supset I(S_2)$ is a proper containment of ideals; if $I \supset I(S)$ is a proper containment of ideals and S is closed then $V(I) \subset S$ is a proper containment of algebraic sets. (These are Exercise 7.2.2.)

The **coordinate ring** of an algebraic set V is the quotient ring

$$K[V] = \mathbf{C}[Z_1,\dots,Z_n]/I(V).$$

Thus an element of $K[V]$, called a **polynomial on** V, is formally a coset $g+I(V)$ for some $g \in \mathbf{C}[Z_1,\dots,Z_n]$, but in practice one just writes g. Polynomials on V describe well-defined functions from V to $\mathbf{C}$. An algebraic set V is a **variety** if its coordinate ring $K[V]$ is an integral domain, or equivalently its ideal $I(V)$ is prime. (Warning: some algebraic geometry texts use "variety" and "irreducible variety" in the respective senses of "algebraic set" and "variety" here.)

A subset S of a topological space X is **dense in** X if its X-closure (i.e., its smallest closed superset in X) is all of X. Equivalently (Exercise 7.2.3(a)), S has nonempty intersection with every nonempty open subset of X.

(7.2.1) LEMMA. *Let V be a variety and let S be a proper closed (algebraic) subset of V. Suppose a polynomial g on V vanishes on all of $V \setminus S$. Then g vanishes on all of V. It follows that $\overline{V \setminus S} = V$, i.e., every nonempty open set is dense in V.*

PROOF. The containment $I(S) \supset I(V)$ is proper. Any polynomial $\tilde{g} \in I(S) \setminus I(V)$ vanishes on S but not on all of V. The product $g\tilde{g}$ thus vanishes on V, i.e., lies in the prime ideal $I(V)$. But $\tilde{g} \notin I(V)$ so the first statement follows and gives $I(V \setminus S) \subset I(V)$. The reverse containment is clear, and taking algebraic sets completes the proof. □

Since a finite intersection of dense open subsets is dense in any topological space (Exercise 7.2.3(b)), a finite union of proper closed subsets of V also has dense open complement. If V is not a variety the lemma fails, e.g., let V be the "axes" $V(\langle XY \rangle)$ in $\mathbf{C}^2$ and let S be the axis $V(\langle X \rangle)$. In developing intuition for the Zariski topology, one should think of proper closed subsets as very small, such as curves on a surface, and of proper open subsets as correspondingly large, such as the full surface with some curves removed. See Exercise 7.2.4 for an example.

The **function field** $K(V)$ of a variety is the field of quotients of its coordinate ring. An element of $K(V)$, called a **rational function on** V, is formally

$$f = \overline{(g, h)} \qquad \text{for some } g, h \in K[V] \text{ with } h \neq 0,$$

where the overbar denotes the quotient field equivalence $\overline{(g,h)} = \overline{(\tilde{g}, \tilde{h})}$ if $g\tilde{h} - \tilde{g}h = 0$ in $K[V]$. Of course, one abbreviates to $f = g/h$ where $g, h \in \mathbf{C}[Z_1, \ldots, Z_n]$ and forgets about cosets and equivalences, but rational functions on V are actually a bit subtle when viewed as mappings. First, the value $g(v)/h(v) \in \mathbf{C}$ is undefined at points $v \in V$ where $h(v) = 0$. This isn't too terrible though: h can only vanish on a proper closed subset of V, which is small in the sense of Lemma 7.2.1. Second, a different representation $f = \tilde{g}/\tilde{h}$ might be defined at different points of V. So, consider f defined at $v \in V$ if *any* of its representatives is defined there; this makes sense because all representatives defined at v take the same value (Exercise 7.2.5(a)). See Exercise 7.2.5(b) for an example of all this.

Let $V_1 \subset \mathbf{C}^n$ and $V_2 \subset \mathbf{C}^m$ be varieties. A vector

$$\phi = (f_1, \ldots, f_m) \qquad \text{with each } f_i \in K(V_1)$$

defines a function from a dense open subset of V_1 to $\mathbf{C}^m$, cf. the first remark after Lemma 7.2.1. If $\phi(v) = (f_1(v), \ldots, f_m(v)) \in V_2$ for each v where all f_i are defined, then ϕ is called a **rational map from V_1 to V_2**. Since ϕ may not be defined on all of V_1, the terminology is misleading, as is the notation $\phi : V_1 \longrightarrow V_2$; some authors use a broken arrow instead. At any rate, a rational map is continuous on its domain (Exercise 7.2.6). In general, the composition of two rational maps may not be defined, e.g., $0 : \mathbf{C} \longrightarrow \mathbf{C}^2$ followed by $X/Y : \mathbf{C}^2 \longrightarrow \mathbf{C}$, but when the composition is defined, it is again rational.

The rational map $\phi : V_1 \longrightarrow V_2$ is **dominant** if the Zariski closure $\overline{\phi(V_1)}$ of its image is all of V_2; for example, at the beginning of the section, the map from the elliptic curve to the w-sphere and the map from polynomial roots to coefficients are dominant, being surjections. For such a map and any rational $\psi : V_2 \longrightarrow V_3$, the composition $\psi \circ \phi$ is defined. Indeed, letting "dom" and "im" denote domain and image, $\phi^{-1}(\text{dom}(\psi))$ is open in $\text{dom}(\phi)$ by continuity; it is nonempty—and therefore dense by Lemma 7.2.1—because the nonempty open set $\text{dom}(\psi)$ and the dense set $\text{im}(\phi)$ intersect. The composition of dominant maps is dominant by three topological facts (Exercise 7.2.7): (1) if S is a dense subset of the topological space X then for any open $Y \subset X$, $S \cap Y$ is dense in Y; (2) if $f : X \longrightarrow Y$ is a continuous map of topological spaces and S is dense in X then $f(S)$ is dense in $f(X)$; and (3) if S is dense in X and S_1 is dense in S then S_1 is dense in X. Granting these, let $\phi : V_1 \longrightarrow V_2$ and $\psi : V_2 \longrightarrow V_3$ be dominant. Then $\text{im}(\phi)$ is dense in V_2, so (1) says $\text{im}(\phi) \cap \text{dom}(\psi)$ is dense in $\text{dom}(\psi)$, (2) says $\psi(\text{im}(\phi) \cap \text{dom}(\psi))$ is dense in $\text{im}(\psi)$, which is dense in V_3, and (3) says $\psi(\text{im}(\phi) \cap \text{dom}(\psi))$ is dense in V_3. This gives the result.

If $\phi : V_1 \longrightarrow V_2$ is rational and there exists another rational map $\psi : V_2 \longrightarrow V_1$ such that $\psi \circ \phi = id_{V_1}$ and $\phi \circ \psi = id_{V_2}$ (i.e., both compositions are defined and the equalities are restricted to suitable dense open sets) then ϕ is a **birational equivalence** from V_1 to V_2. A birational equivalence is necessarily dominant.

Each rational map $\phi : V_1 \longrightarrow V_2$ naturally induces a **pullback** homomorphism $\phi^* : K[V_2] \longrightarrow K(V_1)$ given by

$$\phi^*(g) = g \circ \phi.$$

The pullback of a dominant map is injective, for if $g \circ \phi = 0$ in $K(V_1)$ then $g \in I(\phi(V_1)) = I(\overline{\phi(V_1)}) = I(V_2)$, i.e., $g = 0$ in $K[V_2]$. Thus the pullback of a dominant map extends to an injection of function fields $\phi^* : K(V_2) \longrightarrow K(V_1)$ by $\phi^*(g/h) = (g \circ \phi)/(h \circ \phi)$, also called the pullback. If ϕ is a birational equivalence of varieties then ϕ^* is an isomorphism of function fields (Exercise 7.2.8).

The alert reader has noticed that the algebraic sets defined here differ from those defined in Section 5.6, and in particular that our main motivating example, the Riemann sphere, is not an algebraic set in the present sense. This is easily remedied. Analogously to associating subsets of $\mathbf{C}^n$ to ideals in $\mathbf{C}[Z_1, \ldots, Z_n]$, one can associate subsets of $\mathbf{P}^n(\mathbf{C})$ to **homogeneous ideals** in $\mathbf{C}[Z_1, \ldots, Z_{n+1}]$—these are the ideals that are generated by forms $g \in \mathbf{C}[Z_1 : \cdots : Z_{n+1}]$. The algebraic set associated to such an ideal I generated by forms g_i is

$$V(I) = \{[z_1 : \cdots : z_{n+1}] \in \mathbf{P}^n(\mathbf{C}) : g_i(z_1, \ldots, z_{n+1}) = 0 \text{ for all } i\}.$$

This is a well-defined subset of projective space. Similarly, the homogeneous ideal associated to a subset S of $\mathbf{P}^n(\mathbf{C})$ is

$$I(S) = \langle g \in \mathbf{C}[Z_1 : \cdots : Z_{n+1}] : g(z) = 0 \text{ for all } [z_1 : \cdots : z_{n+1}] \in S \rangle,$$

again well-defined. Now the developments of this section can be repeated in the projective environment with minor changes (these aren't completely trivial; see for example [Ha]). In particular, complex projective varieties have associated function fields. We omit the details.

The right idea of how varieties and their function fields correspond is motivated by the fact that rational functions are generally defined only on dense open subsets of varieties. As far as birational equivalence and function fields are concerned, removing proper subvarieties from a variety is irrelevant. For example, removing the one-point subvariety $V(Z_2)$ from the projective variety $\mathbf{P}^1(\mathbf{C})$ leaves the plane $\mathbf{C}$, which still has field of rational functions $\mathbf{C}(Z)$. Similarly the elliptic curve $\mathcal{E}$ from the beginning of this section is obtained by removing the point $[0 : 1 : 0]$ from its projective counterpart

$$V(\langle Y^2 Z - p^*(X, Z) \rangle)$$

where as in Section 1.4, p^* is the homogenization of p to a cubic form. This projective variety is topologically a complete torus rather than the punctured one from before. To take these sorts of examples into account, introduce the term **quasi-projective variety** to mean an open subset of a variety in $\mathbf{P}^n(\mathbf{C})$.

Now define two sets of equivalence classes. The set of **variety classes** is

$$\mathcal{V} = \{\text{quasi-projective varieties } V\}/\sim$$

where $V_1 \sim V_2$ if V_1 and V_2 are birationally equivalent, so all birationally equivalent quasi-projective varieties are the same object in $\mathcal{V}$. The mappings associated with this set are

$$M_{\mathcal{V}} = \{\text{dominant rational maps } \phi : V_1 \longrightarrow V_2 \text{ of varieties}\}/\sim$$

where $\phi_1 \sim \phi_2$ if $\phi_2 = \psi \circ \phi_1 \circ \theta$ with ψ and θ birational equivalences, so the elements of $M_{\mathcal{V}}$ are naturally viewed as mappings between the elements of $\mathcal{V}$. Thus all dominant rational maps that differ by composition with a birational equivalence at either end are one mapping in $M_{\mathcal{V}}$. Meanwhile, the set of **function field classes** is

$$\mathcal{K} = \{\text{fields } \mathbf{K} \text{ of finite transcendence degree over } \mathbf{C}\}/\sim$$

where $\mathbf{K}_1 \sim \mathbf{K}_2$ if $\mathbf{K}_1$ and $\mathbf{K}_2$ are $\mathbf{C}$-isomorphic. Its associated mappings are

$$M_{\mathcal{K}} = \{\mathbf{C}\text{-injections } \iota : \mathbf{K}_1 \longrightarrow \mathbf{K}_2 \text{ of such fields}\}/\sim$$

where $\iota_1 \sim \iota_2$ if $\iota_2 = \sigma \circ \iota_1 \circ \tau$ with σ and τ $\mathbf{C}$-isomorphisms, so the elements of $M_{\mathcal{K}}$ are naturally viewed as mappings between the elements of $\mathcal{K}$. Thus all $\mathbf{C}$-injections $\iota : \mathbf{K}_1 \longrightarrow \mathbf{K}_2$ that differ by a $\mathbf{C}$-isomorphism at either end are one mapping in $M_{\mathcal{K}}$. This mapping is therefore naturally viewed as the inclusion $\iota(\mathbf{K}_1) \subset \mathbf{K}_2$, or even as the field extension $\mathbf{K}_2/\iota(\mathbf{K}_1)$.

The relation between these two sets encompasses all of our examples so far. See [Ha] for a proof of the following theorem.

(7.2.2) VARIETY–FIELD CORRESPONDENCE. *Taking each quasi-projective variety to its function field,*

$$V \mapsto K(V),$$

gives a bijection between the sets $\mathcal{V}$ *and* $\mathcal{K}$ *of variety classes and function field classes. That is, each field* $\mathbf{K}$ *of finite transcendence degree over* $\mathbf{C}$ *is the function field of some variety, and birationally inequivalent varieties have* $\mathbf{C}$*-nonisomorphic function fields. Taking each dominant rational map of varieties to its pullback,*

$$(\phi : V_1 \longrightarrow V_2) \mapsto (\phi^* : K(V_2) \longrightarrow K(V_1)),$$

give a naturally corresponding bijection between the sets $M_{\mathcal{V}}$ *and* $M_{\mathcal{K}}$ *of mapping classes between varieties and between function fields. That is, each*

$\mathbf{C}$-injection of function fields is the pullback of some dominant rational map between the appropriate varieties, and distinct classes of dominant rational maps pull back to distinct classes of $\mathbf{C}$-injections.

If $\phi : X \longrightarrow V$ is a dominant rational map and the corresponding extension $\mathbf{K}/\mathbf{k} = K(X)/\iota(K(V))$ is Galois, then the pullback gives an anti-isomorphism

$$^* : \mathrm{Cover}(X/V) \longrightarrow \mathrm{Gal}(\mathbf{K}/\mathbf{k}),$$

where $\mathrm{Cover}(X/V)$ *is the* **covering group** *of X over V,*

$$\mathrm{Cover}(X/V) = \{\gamma : X \longrightarrow X : \gamma \textit{ is a birational equivalence and } \phi \circ \gamma = \phi\}.$$

The theorem simply says that at the level of classes, varieties and function fields correspond perfectly, as do dominant maps of varieties and $\mathbf{C}$-injections of function fields. The identification of the groups $\mathrm{Cover}(X/V)$ and $\mathrm{Gal}(\mathbf{K}'/\mathbf{k})$ is precisely the process that was carried out at length in Section 4.4: $\phi : X \longrightarrow V$ was the rotation group invariant $f_\Gamma : \widehat{\mathbf{C}} \longrightarrow \widehat{\mathbf{C}}$ and $\mathrm{Cover}(X/V)$ was the rotation group Γ itself; the covering group condition was $f_\Gamma \circ \gamma = f_\Gamma$, which holds for all $\gamma \in \Gamma$ by nature of the invariant.

We need one more result for future reference. A topological space is **connected** if it is not the union of disjoint nonempty open subsets, and **discrete** if each point is an open set. (Note how this is consistent with our informal use of the word "discrete" in Sections 3.2 and 7.1.) A continuous function from a connected space to a discrete space must be constant (Exercise 7.2.9).

(7.2.3) THEOREM. *Let V be a quasi-projective variety. Then V is a connected subset of $\mathbf{P}^n(\mathbf{C})$ under the usual topology that $\mathbf{P}^n(\mathbf{C})$ inherits from $\mathbf{C}^{n+1}$. In particular, any continuous function from V to a discrete space is constant.*

This is Corollary 4.16 in [Mu].

Exercises

7.2.1. Show that $\emptyset$ and $\mathbf{C}^n$ are algebraic sets. If $V(I_1)$ and $V(I_2)$ are algebraic sets, show that their union is $V(I_1 I_2)$. If $\{V(I_j) : j \in J\}$ is an arbitrary collection of algebraic sets, show that their intersection is $V(\langle I_j : j \in J\rangle)$.

7.2.2. Show that the operators V and I reverse containments, i.e., if $I_1 \subset I_2$ are ideals then $V(I_1) \supset V(I_2)$, and similarly for I. If $S \subset \mathbf{C}^n$ is a set, show that $V(I(S)) \supset S$. If $I \subset \mathbf{C}[Z_1, \ldots, Z_n]$ is an ideal, show that

$I(V(I)) \supset I$. Combine these results to show that $I(V(I(S))) = I(S)$ and $V(I(V(I))) = V(I)$ for sets S and ideals I. Show that $V(I(S))$ is the Zariski closure of S. Show that if $S_1 \subset S_2$ is a proper containment of algebraic sets then $I(S_1) \supset I(S_2)$ is a proper containment of ideals. Show that if $I \supset I(S)$ is a proper containment of ideals and S is closed then $V(I) \subset S$ is a proper containment of algebraic sets.

7.2.3. (a) Show that a subset S of a topological space X is dense if and only if S has nonempty intersection with every nonempty open subset of X.

(b) If X is a topological space and $S_1, \ldots, S_k$ are dense open subsets of X, show that $\bigcap_{i=1}^{k} S_i$ is again a dense open subset of X.

7.2.4. Show that the proper Zariski-closed subsets of $\mathbf{C}$ are the finite sets. (Recall that $\mathbf{C}[Z]$ is a principal ideal domain.)

7.2.5. (a) If V is a variety, $f \in K(V)$ is a rational function on V, and $v \in V$ is a point, show that all representatives of f that are defined at v take the same value there.

(b) Let $I = \langle X^2+Y^2-1 \rangle \subset \mathbf{C}[X,Y]$ and let $V = V(I) \subset \mathbf{C}^2$, a variety. Let $f = (X+1)/Y \in K(V)$. This representation of f is indeterminate at $(-1,0)$ and has a pole at $(1,0)$. Show that also $f = -Y/(X-1)$, so in fact $f(-1,0) = 0$. Can some other representation of f take a finite value at $(1,0)$?

7.2.6. Let $\phi : V_1 \longrightarrow V_2$ be a rational map of varieties, and let $V(I_2)$ be a closed subset of V_2. Define

$$I_1 = \langle g \in K[V_1] : \phi^*(f) = g/h \text{ for some } f \in I_2 \text{ and } h \in K[V_1] \rangle.$$

Show that $\phi^{-1}(V(I_2)) = \operatorname{dom}(\phi) \cap V(I_1)$, so ϕ is continuous on its domain.

7.2.7. Prove the three topological facts cited in the section: if S is a dense subset of X then for any open $Y \subset X$, $S \cap Y$ is dense in Y; if $f : X \longrightarrow Y$ is a continuous map of topological spaces and S is dense in X then $f(S)$ is dense in $f(X)$; and if S is dense in X and S_1 is dense in S then S_1 is dense in X.

7.2.8. If $\phi : V_1 \longrightarrow V_2$ and $\psi : V_2 \longrightarrow V_1$ are rational maps of varieties, show that $(\psi \circ \phi)^* = \phi^* \circ \psi^*$. Show that $id_{V_1}^* = id_{K(V_1)}$. Conclude that if ϕ is a birational equivalence then $\phi^* : K(V_2) \longrightarrow K(V_1)$ is an isomorphism.

7.2.9. Let X be a connected topological space and Y a discrete topological space. Prove that every continuous function $f : X \longrightarrow Y$ is constant.

7.2.10. Use the Variety–Field Correspondence to reinterpret Lüroth's Theorem and the Embedding Lemma from Chapter 6 as results about varieties.

3. Purely iterative algorithms

Varieties and function fields are the natural environment for a general theory of algorithms like Newton's method for nth roots. Throughout this section, let V be a quasi-projective variety with function field $K(V) = \mathbf{k}$. From now on we will just say "variety" and not distinguish among birationally equivalent varieties, i.e., we will really work with variety classes and their corresponding function field classes. This will lead to occasional casual phrasing.

(7.3.1) DEFINITION. *An* **iterative algorithm over a variety** V *with function field* $\mathbf{k}$ *(or* **over a function field** $\mathbf{k}$ *with variety* V*) is an element of the field* $\mathbf{k}(T)$ *of degree* $d > 1$*. Equivalently, an iterative algorithm over* V *is a rational map from* V *to the set of complex rational functions in* T *of degree* d*,*

$$F : V \longrightarrow \mathbf{C}(T), \qquad v \mapsto F_v.$$

The algorithm associated to each $v \in V$ *is to pick some initial guess* $t \in \widehat{\mathbf{C}}$ *and iterate* F_v *on* t*, obtaining the sequence*

$$\{t, F_v(t), F_v(F_v(t)), \ldots\} = \{F_v^k(t) : k = 0, 1, 2, \ldots\}.$$

For example, Newton's method for nth roots from Section 7.1, $F = ((n-1)T^n + W)/(nT^{n-1})$ (or equivalently $F : w \mapsto ((n-1)T^n + w)/(nT^{n-1})$), is an iterative algorithm over the variety $\mathbf{C}$ with function field $\mathbf{C}(W)$.

The iteration associated to F is decision-free. To emphasize this property, the algorithms in Definition 7.3.1 are also called **purely** iterative algorithms. Since F_v is undefined on a proper closed subset of V, the "each $v \in V$" in the definition is an instance of phrasing to be parsed at the level of variety classes—it makes sense once we remove the proper closed subset of bad points from V. In the case of Newton's method for nth roots, for example, the point $w = 0$ must be removed from $\mathbf{C}$ since $F_0 = ((n-1)/n)T$ does not have requisite degree $d > 1$.

In Section 7.1, an output point for Newton's method was a point $(w, z) \in \widehat{\mathbf{C}} \times \widehat{\mathbf{C}}$ such that $z^n = w$, or equivalently $F_w(z) = z$. Analogously, it is tempting to define an output point of any iterative algorithm F to be a point $(v, z) \in V \times \widehat{\mathbf{C}}$ such that $F_v(z) = z$. This idea is a little too simple, though. A key feature of Newton's method is that any iteration starting close to an nth root z of w will in fact converge to z. Thus the appropriate definition is

(7.3.2) DEFINITION. *For any $f \in \mathbf{C}(T)$, the* **attractor** *of f is the set*

$$\mathrm{Att}(f) = \{z \in \widehat{\mathbf{C}} : z = \lim_{k\to\infty} f^k(t) \text{ for all } t \text{ in some neighborhood } U_z \text{ of } z\}.$$

The **output** *of an iterative algorithm F is*

$$\mathrm{Out}(F) = \{(v, z) \in V \times \widehat{\mathbf{C}} : z \in \mathrm{Att}(F_v)\}.$$

The attractor of any $f = g/h \in \mathbf{C}(T)$ is a subset of the fixed points of f, which in turn are the finite algebraic set $V(\langle g - hT\rangle)$ (Exercise 7.3.1). Similarly, the output of an iterative algorithm $F = G/H \in \mathbf{k}(T)$ is a subset of the algebraic set $V(\langle G - HT\rangle)$. An iterative algorithm whose output is a variety (again, at the level of classes) is called **simple**. We will only discuss simple algorithms since they are all we need and the general case is slightly more technical.

(7.3.3) DEFINITION. *An iterative algorithm $F \in \mathbf{k}(T)$ over V is* **generally convergent** *if for each v in a dense open subset of V, and for each t in a dense open subset D_v of initial guesses in $\widehat{\mathbf{C}}$, the sequence*

$$\{F_v^k(t) : k = 0, 1, 2, \dots\}$$

obtained by iterating F_v on t converges to some $z \in \mathrm{Att}(F_v)$. For a given v, different initial guesses $t \in D_v$ may iterate under F_v to different attracting points z.

One may picture the output of an iterative algorithm F as the graph of a multiple-valued function from V to $\widehat{\mathbf{C}}$, with the points of the attractor $\mathrm{Att}(F_v)$ forming the cross-section over each point $v \in V$. Iterating a generally convergent algorithm moves initial points (v, t) vertically to points (v, z) on the graph. This was all depicted in Figure 7.1.1 for Newton's method; the only difference now is that the horizontal axis more generally consists of points v from the variety V rather than points w from the sphere. Of course, not all algorithms are generally convergent. We will have many counterexamples after the next theorem.

Let F be a simple algorithm over the variety V with function field $\mathbf{k}$. Computationally, $\mathbf{k}$ is the set of algorithms that take input from V and perform finitely many rational operations. The larger function field $\mathbf{K} = K(\mathrm{Out}(F))$ is similarly the algorithms that perform finitely many rational operations on the output variety $\mathrm{Out}(F)$; this is the **output field** of F. If F is generally convergent then it reliably computes $\mathrm{Out}(F)$, so the elements of $\mathbf{K}$ are the algorithms that take input from V, iterate F, and then perform finitely many rational operations. That is, the generally convergent simple algorithm F constructs the function field extension $\mathbf{K}/\mathbf{k}$ in a computational sense.

We have seen all this when F is Newton's method for nth roots. It is iterative over the w-sphere $V = \widehat{\mathbf{C}}$ with function field $\mathbf{k} = \mathbf{C}(W)$. For each $w \in \widehat{\mathbf{C}} \setminus \{0, \infty\}$, the attractor $\mathrm{Att}(F_w)$ is $\{z \in \mathbf{C} : z^n = w\}$, so the output is

$$\mathrm{Out}(F) = \{(w, z) \in \mathbf{C} \times \mathbf{C} : z^n = w\}.$$

This is the z-sphere as a variety class (Exercise 7.3.2). Thus F is simple with output variety $\mathrm{Out}(F) = \widehat{\mathbf{C}}$ and corresponding output field as before,

$$\mathbf{K} = \mathbf{C}(W, Z)/\langle Z^n - W\rangle = \mathbf{C}(Z) \quad \text{where } Z^n = W.$$

Since F is generally convergent, it constructs the extension $\mathbf{C}(Z)/\mathbf{C}(W)$.

The next theorem ([Do-Mc], Theorem 3.1) describes all generally convergent algorithms. It follows from a general theorem on dynamics in [Mc] whose proof lies beyond the scope of this book.

(7.3.4) THEOREM. *Let $F \in \mathbf{k}(T)$ be a simple generally convergent algorithm over V with output variety $\mathrm{Out}(F)$ and output field $\mathbf{K}$. Let $\mathbf{K}'$ be the Galois closure of $\mathbf{K}$ over $\mathbf{k}$. Then F is a composition of functions*

$$F = \phi^{-1} f \phi.$$

Here $f \in \mathbf{C}(T)$, and $\phi = \begin{bmatrix} a & b \\ c & d \end{bmatrix} \in \mathrm{PGL}_2(\mathbf{K}')$ is viewed as the fractional linear transformation $(aT + b)/(cT + d) \in \mathbf{K}'(T)$.

We have seen that Newton's method for nth roots conforms to Theorem 7.3.4 with $\mathbf{K}' = \mathbf{K} = \mathbf{C}(Z)$.

The Variety–Field Correspondence gives a geometric interpretation of the theorem. Let X be the variety corresponding to $\mathbf{K}'$ and let $\pi : X \longrightarrow V$ be the dominant rational map corresponding to the function field extension $\mathbf{K}'/\mathbf{k}$. Then ϕ is an X-parametrized family of fractional linear transformations of T and f is a rational function of T, while the composite $F = \phi^{-1} f \phi$ is by hypothesis a V-parametrized family of rational functions of T. For each $v \in V$, the relation in the theorem specializes to $F_v = \phi_x^{-1} f \phi_x$ in $\mathbf{C}(T)$, where x is any point in X lying over v (i.e., $x \in \pi^{-1}(v)$) and

$\phi_x = \begin{bmatrix} a(x) & b(x) \\ c(x) & d(x) \end{bmatrix} \in \mathrm{PGL}_2(\mathbf{C})$; this shows that each function F_v to iterate is conjugate to the single v-independent function f. As before, the family $\{F_v : v \in V\}$ is a **rigid family** of rational maps, the v-independent f is the **model** for F, and ϕ is the **conjugating conformal map**. In this language, Theorem 7.3.4 asserts that

a generally convergent algorithm is a rigid family of rational maps.

The values that can be computed by a generally convergent algorithm F are severely constrained by rigidity. As Exercise 7.3.4 shows, all cross-sections of $\mathrm{Out}(F)$ for rigid F must be conformally equivalent. For example, no algorithm can be generally convergent to the roots of the general polynomial p of degree $n \geq 4$ since most n-tuples of complex numbers are not conformally equivalent. (Recall, as mentioned in Section 7.1, that the group $\mathrm{PGL}_2(\mathbf{C})$ acts triply transitively on the sphere $\widehat{\mathbf{C}}$.) In particular, applying Newton's method to p does not give a generally convergent algorithm. A generic cubic polynomial, on the other hand, has three complex roots, and all triples in $\widehat{\mathbf{C}}$ are conformally equivalent. So a simple algorithm might compute the roots of the general cubic, and in fact, [Do-Mc] gives such an algorithm: to find the roots of the depressed cubic $p = T^3 + bT + c$, apply Newton's method to the rational function

$$\frac{T^3 + bT + c}{3bT^2 + 9cT - b^2}.$$

Applying Newton's method to the depressed cubic p itself turns out not to be generally convergent.

Next we derive some consequences of rigidity. These consequences will show (in the first part of the pending Theorem 7.3.5) that along with constraining the values that can be computed by a generally convergent algorithm, rigidity also constrains the function field extensions that can be constructed by such an algorithm.

The consequences of rigidity that we need in the general case are the same as those we derived for Newton's method for nth roots in Section 7.1. The conjugating map ϕ lies in $\mathbf{K}'(T)$, but the composition $F = \phi^{-1} f \phi$ lies in $\mathbf{k}(T)$, making it invariant under the Galois group $\Gamma = \mathrm{Gal}(\mathbf{K}'/\mathbf{k})$; therefore (Exercise 7.3.3) for each $\gamma \in \Gamma$, the fractional linear transformation

$$\rho(\gamma) = \phi^\gamma \phi^{-1} = \begin{bmatrix} a^\gamma & b^\gamma \\ c^\gamma & d^\gamma \end{bmatrix} \begin{bmatrix} a & b \\ c & d \end{bmatrix}^{-1} \in \mathrm{PGL}_2(\mathbf{K}')$$

commutes with f.

To show that $\rho(\gamma)$ actually lies in $\mathrm{PGL}_2(\mathbf{C})$, again let $\pi : X \longrightarrow V$ be the dominant rational map corresponding to the extension $\mathbf{K}'/\mathbf{k}$, and let $x \in X$. It suffices to show that the transformation

$$\rho(\gamma)_x = \phi_x^\gamma \phi_x^{-1} = \begin{bmatrix} a^\gamma(x) & b^\gamma(x) \\ c^\gamma(x) & d^\gamma(x) \end{bmatrix} \begin{bmatrix} a(x) & b(x) \\ c(x) & d(x) \end{bmatrix}^{-1} \in \mathrm{PGL}_2(\mathbf{C})$$

is independent of x. By Exercise 7.3.4, $\rho(\gamma)_x$ permutes the attractor $A = \mathrm{Att}(f)$. But A is finite, so the set $\mathrm{Aut}(A)$ of fractional linear transformations that permute it is also finite (Exercise 7.3.5); since finite sets are discrete, the continuous map $X \longrightarrow \mathrm{Aut}(A)$ taking $x \mapsto \rho(\gamma)_x$ is constant by Theorem 7.2.3.

It follows that $\rho : \Gamma \longrightarrow \mathrm{PGL}_2(\mathbf{C})$ is a homomorphism precisely as in Section 7.1 (Exercise 7.3.6). Also, ρ is injective by the following argument. Assume that the output field extension is Galois, i.e., $\mathbf{K}' = \mathbf{K}$ and $X = \mathrm{Out}(F)$; the non-Galois case is more technical, so we skip it for clarity. Under this assumption, each $x \in X$ takes the form $x = (v, z) \in V \times \widehat{\mathbf{C}}$ with $z \in \mathrm{Att}(F_v)$, and π is vertical projection from the multiple-valued graph X to V. (Drawing a picture of this configuration and a separate vertical $\widehat{\mathbf{C}}$ containing A may help with the rest of this paragraph.) Define a function $\alpha : X \longrightarrow \widehat{\mathbf{C}}$ by

$$\alpha(x) = \phi_x(z).$$

Exercise 7.3.4 shows that for each $x = (v, z)$, the transformation ϕ_x maps the set of output values $\mathrm{Att}(F_v)$ bijectively to A, so α maps X to the finite set A and is therefore constant by Theorem 7.2.3. Recall from the Variety–Field Correspondence that Γ may be viewed as the covering group of X/V. With this identification, for any $\gamma \in \Gamma$ and $x \in X$,

$$\phi_x^\gamma = \begin{bmatrix} a^\gamma(x) & b^\gamma(x) \\ c^\gamma(x) & d^\gamma(x) \end{bmatrix} = \begin{bmatrix} a(\gamma x) & b(\gamma x) \\ c(\gamma x) & d(\gamma x) \end{bmatrix} = \phi_{\gamma x}.$$

Now let γ be a nonidentity element of Γ. For some $v \in V$, γ permutes the cross-section $\pi^{-1}(v) = v \times \mathrm{Att}(F_v)$ nontrivially; that is, for some $z \in \mathrm{Att}(F_v)$, $\gamma x = \gamma(v, z) = (v, z')$ with $z' \neq z$. So γx has the same v-coordinate as x, and $\phi_{\gamma x}$ also maps $\mathrm{Att}(F_v)$ bijectively to A. Since α is constant,

$$\phi_{\gamma x}(z) \neq \phi_{\gamma x}(z') = \alpha(\gamma x) = \alpha(x) = \phi_x(z).$$

Therefore

$$\rho(\gamma)(\phi_x(z)) = (\phi_x^\gamma \phi_x^{-1})(\phi_x(z)) = \phi_{\gamma x}(z) \neq \phi_x(z),$$

showing that $\rho(\gamma)$ is a nontrivial permutation of A. Incidentally, the identity $\rho(\gamma) = \phi_x^\gamma \phi_x^{-1}$ now shows that in fact γ permutes all cross-sections of X as $\rho(\gamma)$ permutes A.

This analysis gives a refinement of Theorem 7.3.4 and—up to some more omitted technicalities—a converse.

(7.3.5) THEOREM. *Let V be a variety with function field $\mathbf{k}$. Let F be a simple generally convergent algorithm over $\mathbf{k}$ with output field $\mathbf{K}$. Let $\mathbf{K}'$ be the Galois closure of $\mathbf{K}$ over $\mathbf{k}$ with* $\mathrm{Gal}(\mathbf{K}'/\mathbf{k}) = \Gamma$. *Then F takes the form*

$$F = \phi^{-1} f \phi$$

with $f \in \mathbf{C}(T)$ and $\phi \in \mathrm{PGL}_2(\mathbf{K}')$. For such F, the model f produces a convergent iteration to its attractor A on a dense open subset D of $\widehat{\mathbf{C}}$. The map

$$\rho : \gamma \mapsto \phi^\gamma \phi^{-1}$$

is an embedding $\Gamma \longrightarrow \mathrm{Aut}(f)$, *where*

$$\mathrm{Aut}(f) = \{m \in \mathrm{PGL}_2(\mathbf{C}) : m^{-1} f m = f\}.$$

Conversely, let V again be a variety with function field $\mathbf{k}$. Suppose $\mathbf{K}'/\mathbf{k}$ is a Galois extension with Galois group Γ, and we have

1. *a model $f \in \mathbf{C}(T)$ and a finite attractor $A \subset \widehat{\mathbf{C}}$ such that for each t in a dense open subset D of $\widehat{\mathbf{C}}$, the iteration $\{f^k(t) : k = 0, 1, 2, \dots\}$ converges to a point $a \in A$,*
2. *a transformation $\phi \in \mathrm{PGL}_2(\mathbf{K}')$ such that the map $\gamma \mapsto \phi^\gamma \phi^{-1}$ is an embedding $\rho : \Gamma \longrightarrow \mathrm{Aut}(f)$ and $\rho(\Gamma)$ acts transitively on A. ($\rho(\Gamma)$ is a subgroup of $\mathrm{PGL}_2(\mathbf{C})$.)*

Then the composition $F = \phi^{-1} f \phi$ is a simple generally convergent algorithm over $\mathbf{k}$. Its output field is $\mathbf{k}(\phi^{-1}(a))$ for any $a \in A$, which is the fixed field in $\mathbf{K}'$ of $\rho^{-1}(\mathrm{stab}(a))$. ($\rho^{-1}(\mathrm{stab}(a))$ is a subgroup of Γ.)

As mentioned above, the function field extensions that can be constructed by a simple algorithm F are constrained by the first half of Theorem 7.3.5. Specifically, the embedding $\Gamma \longrightarrow \mathrm{Aut}(f) \subset \mathrm{PGL}_2(\mathbf{C})$ in Theorem 7.3.5 shows that Γ is isomorphic to a rotation group of the sphere, so by Theorem 2.6.1, Γ must be cyclic, dihedral or Platonic. Thus we now know for the general polynomial p of any degree $n \geq 5$ that no generally convergent algorithm F over the coefficient field can construct a superfield of the root field inside the splitting field: the Galois group S_n is prohibitive. This result is stronger than our earlier observation that F can not converge to the roots of p, since the roots might still be rational functions of the output of F. For $n = 4$, the general quartic has Platonic group S_4, so an algorithm could conceivably construct a superfield of the quartic root field, but in fact, [Do-Mc] shows that no such algorithm exists either.

Most importantly for us, since the Brioschi quintic has Platonic group A_5, an algorithm might construct a superfield of its root field inside its splitting field. This will turn out to be the case.

The second half of Theorem 7.3.5 complements Theorem 7.3.4 by asserting that

a rigid family of rational maps satisfying certain conditions

is a generally convergent algorithm.

In other words, finding a suitable model and conjugating map now guarantees a generally convergent algorithm.

Finally, a technical remark on the second half of the theorem: the requirement that $\rho(\Gamma)$ act transitively on A is what makes $\mathrm{Out}(F)$ a variety so that F is simple. The transitivity also makes all fixed fields of $\rho^{-1}(\mathrm{stab}(a))$ $\mathbf{C}$-isomorphic as a varies through A, so the output field described makes sense as a function field class (Exercise 7.3.7). The identity $(\phi^{-1})^{\gamma}(a) = \phi^{-1}(\rho(\gamma^{-1})a)$, which follows from the definition $\rho(\gamma) = \phi^{\gamma}\phi^{-1}$ (Exercise 7.3.8(a)), shows that $\mathbf{k}(\phi^{-1}(a))$ is indeed the fixed field in $\mathbf{K}'$ of $\rho^{-1}(\mathrm{stab}(a))$ (Exercise 7.3.8(b)). As explained in Exercise 7.3.4, ϕ^{-1} bijects A to the various sections of $\mathrm{Out}(F)$, so it really does compute the outputs of F as claimed.

Exercises

7.3.1. Show that the attractor of any $f = g/h \in \mathbf{C}(T)$ is a subset of the fixed points of f, which in turn are the finite algebraic set $V(\langle g - hT\rangle)$.

7.3.2. Show that the set $\{(w, z) \in \mathbf{C} \times \mathbf{C} : z^n = w\}$ is equivalent to the projective line $\mathbf{P}^1(\mathbf{C})$ as a variety class.

7.3.3. Show that the fractional linear transformation $\rho(\gamma) = \phi^{\gamma}\phi^{-1}$ commutes with f.

7.3.4. In the notation of the section, let $v = \pi(x)$. Use the relation $f = \phi F \phi^{-1}$ to show that for any $z \in \widehat{\mathbf{C}}$, the following two conditions are equivalent:

$$z = \lim_k F_v^k(t) \text{ for all } t \text{ in some neighborhood } U_z \text{ of } z,$$

$$\phi_x(z) = \lim_k f^k(t) \text{ for all } t \text{ in some neighborhood } U_{\phi_x(z)} \text{ of } \phi_x(z).$$

Thus ϕ_x bijects $\mathrm{Att}(F_v)$ to $A = \mathrm{Att}(f)$. This shows that all cross-sections of $\mathrm{Out}(F)$ are conformally equivalent for any rigid F. Now let $\gamma \in \Gamma$ and use the relation $f = \phi^\gamma F(\phi^\gamma)^{-1}$ (see the hint to Exercise 7.1.5) to show similarly that ϕ_x^γ also bijects $\mathrm{Att}(F_v)$ to A. It follows that $\rho(\gamma)_x$ permutes A. This can also be proved directly from the relation $\rho(\gamma)f = f\rho(\gamma)$.

7.3.5. Let A be a finite subset of $\widehat{\mathbf{C}}$ with $|A| \geq 3$. Show that $\mathrm{Aut}(A)$, the fractional linear transformations that permute A, are a finite—and therefore discrete—set. (Recall again that fractional linear transformations act triply transitively on $\widehat{\mathbf{C}}$.) Thus the text's proof that $\rho(\gamma)$ lies in $\mathrm{PGL}_2(\mathbf{C})$ tacitly assumes $|\mathrm{Att}(f)| \geq 3$. This will hold in our application.

7.3.6. Show that $\rho : \Gamma \longrightarrow \mathrm{PGL}_2(\mathbf{C})$ is a homomorphism.

7.3.7. Show that the function field described by the converse in Theorem 7.3.5 is well-defined at the level of function field classes.

7.3.8. (a) Prove the identity $(\phi^{-1})^\gamma(a) = \phi^{-1}(\rho(\gamma^{-1})a)$ as follows. Start from the definition $\rho(\gamma) = \phi^\gamma\phi^{-1}$, take inverses, recall that ρ is a homomorphism and see the hint to Exercise 7.1.5, left-multiply by ϕ^{-1}, and apply both sides to a.

(b) Use the identity to show that $\mathbf{k}(\phi^{-1}(a))$ is indeed the fixed field in $\mathbf{K}'$ of $\rho^{-1}(\mathrm{stab}(a))$.

4. Iteratively constructible extensions

Just as the Radical Criterion describes extensions that can be constructed by successively adjoining radicals, we are now interested in extensions that can be constructed by successively applying simple generally convergent algorithms. Investigating such extensions requires some obvious parallels to the classical terminology from Section 4.5.

(7.4.1) DEFINITION. *Let* $\mathbf{k}$ *be a function field. An* **iteration tower over** $\mathbf{k}$ *is a sequence of fields* $\mathbf{k} \subset \mathbf{k}_1 \subset \mathbf{k}_2 \subset \cdots \subset \mathbf{k}_d$ *such that*

$\mathbf{k}_1$ *is the output field of a simple generally convergent algorithm* F_1 *over* $\mathbf{k}$,

$\mathbf{k}_2$ *is the output field such an algorithm* F_2 *over* $\mathbf{k}_1$,

$\vdots$

$\mathbf{k}_d$ *is the output field of such an algorithm* F_d *over* $\mathbf{k}_{d-1}$.

An extension $\mathbf{K}/\mathbf{k}$ *of function fields is* **constructible by iteration** *if there exists an iteration tower* $\mathbf{k} \subset \cdots \subset \mathbf{k}_d$ *with* $\mathbf{K} \subset \mathbf{k}_d$*; or equivalently,* *if there exists a succession $F_1, \ldots, F_d$ of simple generally convergent algorithms, starting over* $\mathbf{k}$ *and each defined over the output field of its predecessor, whose final output field contains* $\mathbf{K}$.

While the classical algebraic adjunction of an nth root constructs a cyclic extension, a simple iterative algorithm constructs an extension whose group is isomorphic to a rotation group of the sphere. So define a **Möbius group** to be any group isomorphic to one of C_n, D_n, A_4, S_4, or A_5. Again in parallel to the classical language,

(7.4.2) DEFINITION. *A finite group* Γ *is* **nearly solvable** *if there exists a chain of subgroups*

$$\{1\} = \Gamma_d \lhd \Gamma_{d-1} \lhd \cdots \lhd \Gamma_1 \lhd \Gamma_0 = \Gamma$$

each normal in the next and with each quotient Γ_i/Γ_{i-1} *a Möbius group. As in the case of solvable groups, this chain is called a* **subnormal series** *for* Γ.

This definition is unaffected if each quotient is stipulated to be either cyclic or the alternating group A_5 (Exercise 7.4.1). Any subgroup or quotient group of a nearly solvable group is again nearly solvable (Exercise 7.4.2).

The analog to the Radical Criterion is the

(7.4.3) ITERATION CRITERION (OVER $\mathbf{C}$). *Let* $\mathbf{k}$ *be a function field, let* $\mathbf{K}/\mathbf{k}$ *be a finite extension, and let* $\mathbf{K}'$ *be the Galois closure of* $\mathbf{K}$ *over* $\mathbf{k}$. *Then*

$$\mathbf{K}/\mathbf{k} \text{ is constructible by iteration} \iff \mathrm{Gal}(\mathbf{K}'/\mathbf{k}) \text{ is nearly solvable.}$$

Granting this, the classical argument that the general polynomial extension of degree n is constructible by radicals only for $n \leq 4$ now carries over to show that the extension is contructible by iteration for $n \leq 5$. In other words, iteration is one degree stronger than radicals. Specializing to $n = 4$ does not contradict the earlier claim (from the preceding section) that no simple algorithm constructs the quartic root extension: constructing the extension requires a succession of several algorithms rather than merely one.

PROOF. ($\Longrightarrow$) Suppose $\mathbf{K}/\mathbf{k}$ is constructible by iteration. Take an iteration tower $\mathbf{k} = \mathbf{k}_0 \subset \mathbf{k}_1 \subset \cdots \subset \mathbf{k}_d$ with $\mathbf{K} \subset \mathbf{k}_d$. Define $\mathbf{k}'_0 = \mathbf{k}_0$, and for $i = 1, \ldots, d$

$$\begin{aligned} \mathbf{l}_i &= \text{ the Galois closure of } \mathbf{k}_i \text{ over } \mathbf{k}_{i-1}, \\ \mathbf{k}'_i &= \text{ the Galois closure of } \mathbf{k}_i \text{ over } \mathbf{k}. \end{aligned}$$

Exercise 7.4.3(a) shows that we have a tower $\mathbf{k}'_0 \subset \mathbf{k}'_1 \subset \cdots \subset \mathbf{k}'_d$ in which each extension $\mathbf{k}'_i/\mathbf{k}'_{i-1}$ is Galois, as is the net extension $\mathbf{k}'_d/\mathbf{k}'_0 = \mathbf{k}'_d/\mathbf{k}$ by definition of $\mathbf{k}'_d$. If Γ_i denotes the group corresponding to the intermediate field $\mathbf{k}'_i$ in the tower then the group for each step up the tower is the quotient $\mathrm{Gal}(\mathbf{k}'_i/\mathbf{k}'_{i-1}) = \Gamma_{i-1}/\Gamma_i$, denoted Q_i. See Figure 7.4.1.

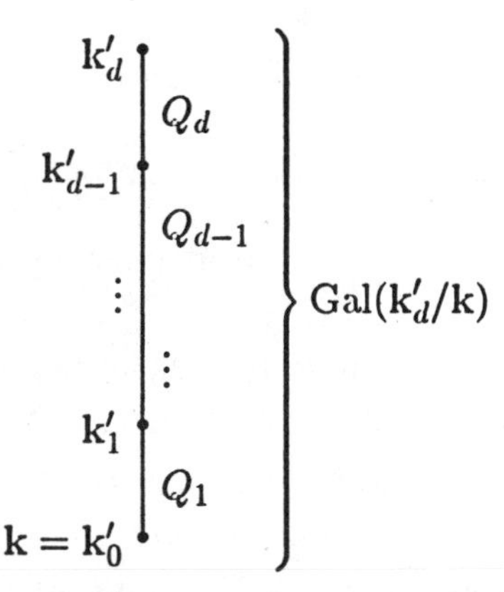

Figure 7.4.1. Galois extensions and corresponding quotient groups

For $i = 1, \ldots, d$, the containments $\mathbf{k}_i \subset \mathbf{l}_i \subset \mathbf{k}'_i$ (Exercise 7.4.3(b)) give the situation shown in Figure 7.4.2. As in Proposition 5.5.1, Q_i injects into $\mathrm{Gal}(\mathbf{l}_i/\mathbf{k}_{i-1})$. Since this last group is Möbius by the first part of Theorem 7.3.5, so is Q_i. Thus the subnormal series

$$\{1\} = \Gamma_d \lhd \Gamma_{d-1} \lhd \cdots \lhd \Gamma_1 \lhd \Gamma_0 = \mathrm{Gal}(\mathbf{k}'_d/\mathbf{k})$$

shows that $\mathrm{Gal}(\mathbf{k}'_d/\mathbf{k})$ is nearly solvable. Since $\mathbf{K}'$ lies in $\mathbf{k}'_d$, $\mathrm{Gal}(\mathbf{K}'/\mathbf{k})$ is a quotient of $\mathrm{Gal}(\mathbf{k}'_d/\mathbf{k})$ and is therefore nearly solvable as well. This completes the proof in one direction.

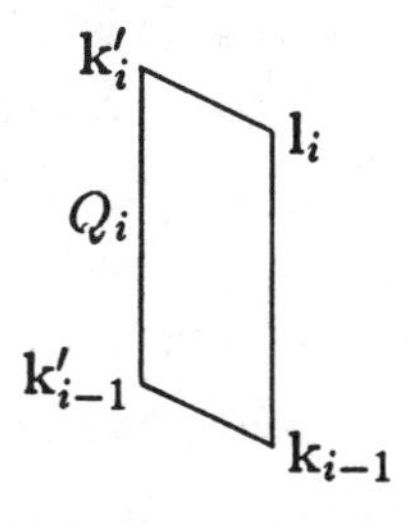

Figure 7.4.2. Q_i injects into a Möbius group

($\Longleftarrow$) Now suppose $\mathrm{Gal}(\mathbf{K}'/\mathbf{k})$ is nearly solvable. The Galois Correspondence gives Figure 7.4.3.

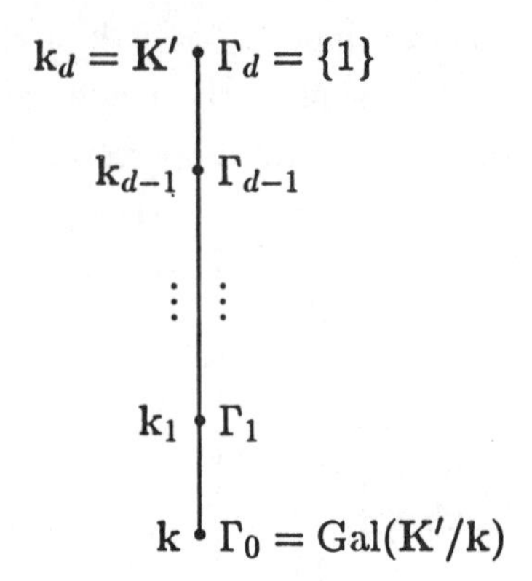

Figure 7.4.3. Subnormal series and corresponding tower of fields

Thus $\Gamma_1 \lhd \Gamma_0$ and by the first remark after Definition 7.4.2, Γ_0/Γ_1 is without loss of generality either a cyclic group C_n or the alternating group A_5. In the cyclic case, the Lagrange Lemma shows that $\mathbf{k}_1 = \mathbf{k}(Z)$ with $Z^n \in \mathbf{k}$. Let $W = Z^n$. Newton's method for nth roots constructs the extension $\mathbf{C}(Z)/\mathbf{C}(W)$, so since $\mathbf{k}_1 = \mathbf{k}(Z)$, Newton's method also constructs the extension $\mathbf{k}_1/\mathbf{k}$ in the tower of fields. The proof can now proceed up the tower for $\mathbf{k}_d/\mathbf{k}_1$, whose group is a subgroup of $\mathrm{Gal}(\mathbf{K}'/\mathbf{k})$.

In the alternating case, let $\mathbf{l}$ be the fixed field of the subgroup A_4 of the Galois group A_5. (See Figure 7.4.4 for a diagram of the objects in this paragraph.) Since A_4 is not normal in A_5 the extension $\mathbf{l}/\mathbf{k}$ is not Galois, but in any case, $[\mathbf{l} : \mathbf{k}] = 60/12 = 5$, so any $Z \in \mathbf{l} \setminus \mathbf{k}$ satisfies a quintic polynomial p over $\mathbf{k}$. Adjoining an auxiliary quadratic irrationality if necessary gives a field $\mathbf{k}'$ over which the quintic p transforms to Brioschi form b. The remainder of this chapter will produce an iterative algorithm that finds a root of b. Granting such an algorithm, the corresponding root Z of p can then be found reliably by radicals, i.e., by further iteration. Let $\mathbf{l}' = \mathbf{k}'\mathbf{l} = \mathbf{k}'(Z)$; thus the extension $\mathbf{l}'/\mathbf{k}$ is constructible by a succession of algorithms. Finally, let $\mathbf{k}_1' = \mathbf{k}_1\mathbf{l}'$. By Proposition 5.5.1, the Galois group of the extension $\mathbf{k}_1'/\mathbf{l}'$ embeds in A_4, so the extension can be constructed by iteration as argued above. Putting all this together shows that iteration constructs the extension $\mathbf{k}_1'/\mathbf{k}$. Since the Galois group $\mathrm{Gal}(\mathbf{K}\mathbf{k}_1'/\mathbf{k}_1')$ is a subgoup of $\mathrm{Gal}(\mathbf{K}/\mathbf{k}_1)$, the proof can continue up a new tower with top field $\mathbf{k}_d' = \mathbf{k}_d\mathbf{k}_1'$. □

Thus the remaining order of business is to solve the Brioschi quintic by iteration. We will do so after developing some general results in the next two sections.

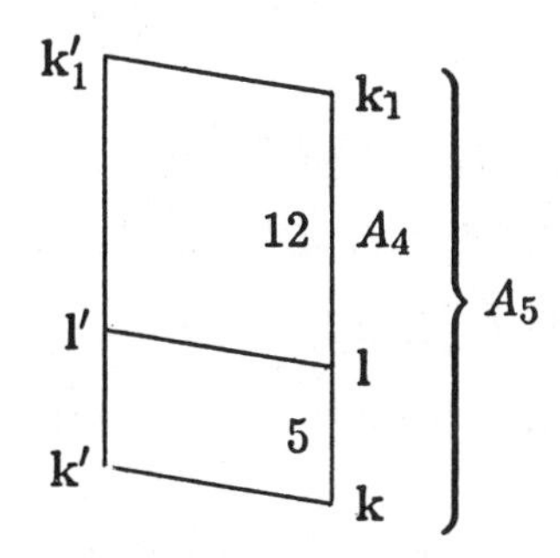

Figure 7.4.4. Constructing an A_5 extension

Exercises

7.4.1. Show that Definition 7.4.2 is unaltered if each quotient is stipulated to be either cyclic or the alternating group A_5.

7.4.2. Show that any subgroup or quotient group of a nearly solvable group is again nearly solvable. (See Exercise 4.5.2 if necessary.)

7.4.3. (a) In the proof of Theorem 7.4.3 ($\Longrightarrow$), explain why $\mathbf{k}'_{i-1} \subset \mathbf{k}'_i$ for $i = 1, \ldots, d$ giving a tower as claimed. Explain why each extension $\mathbf{k}'_i/\mathbf{k}'_{i-1}$ is Galois. Letting Γ_i denote the group corresponding to the intermediate field $\mathbf{k}'_i$ in the tower, explain why the group for each step up the tower is the quotient Γ_{i-1}/Γ_i.

(b) In the same proof, explain the containments $\mathbf{k}_i \subset \mathbf{l}_i \subset \mathbf{k}'_i$ for $i = 1, \ldots, d$.

5. Differential forms

An iterative algorithm needs as its model a rational function f that commutes with a finite subgroup of $\mathrm{PGL}_2(\mathbf{C})$. The next section will construct such functions with the help of **differential forms** on $\mathbf{P}^1(\mathbf{C})$, the subject of this section. Differential forms are defined as follows: a **0-form** is simply any element of $F \in \mathbf{C}[Z_1 : Z_2]$, i.e., a form in the usual sense. A **1-form** is any element

$$F_1 dZ_1 + F_2 dZ_2 \qquad \text{with } F_1 \text{ and } F_2 \text{ 0-forms of the same degree.}$$

Here dZ_1 and dZ_2 are **differentials**, new symbols whose manipulation rules will be given in a moment. Finally, a **2-form** is any element $F dZ_1 dZ_2$ where F is a 0-form. Forms are multiplied subject to all the algebraic rules one would expect, and the additional **skew symmetry** rule for differentials: $dZ_j dZ_i = -dZ_i dZ_j$ for all $i, j \in \{1, 2\}$. Thus $dZ_1 dZ_1 = dZ_2 dZ_2 = 0$ and

$dZ_2 dZ_1 = -dZ_1 dZ_2$ (Exercise 7.5.1). For some more general examples,

(7.5.1)
$$\begin{aligned} F(F_1 dZ_1 + F_2 dZ_2) &= (FF_1)dZ_1 + (FF_2)dZ_2, \\ (F_1 dZ_1 + F_2 dZ_2)(G_1 dZ_1 + G_2 dZ_2) &= (F_1 G_2 - F_2 G_1)dZ_1 dZ_2, \\ (F_1 dZ_1 + F_2 dZ_2)F dZ_1 dZ_2 &= 0. \end{aligned}$$

The **differentiation operator** d takes 0-forms to 1-forms, 1-forms to 2-forms, and 2-forms to 0. Letting D_1 and D_2 denote partial differentiation with respect to Z_1 and Z_2, the differentiation rules are

$$\begin{aligned} dF &= (D_1 F)dZ_1 + (D_2 F)dZ_2, \\ d(F_1 dZ_1 + F_2 dZ_2) &= (D_1 F_2 - D_2 F_1)dZ_1 dZ_2, \\ d(F dZ_1 dZ_2) &= 0. \end{aligned}$$

Note that as a special case, $d(Z_i) = dZ_i$ for $i = 1, 2$, so all of this notation is consistent. Applying d twice in succession to a 0-form gives the 2-form 0, i.e., $d^2 = 0$. Equivalently, if a 1-form μ takes the form $\mu = dF$ for some 0-form F, then $d\mu = 0$. The converse statement is also true: if $d\mu = 0$ for a 1-form μ, then $\mu = dF$ for some 0-form F. (Exercise 7.5.2 asks for proofs of these statements.)

Each $\gamma' \in \mathrm{GL}_2(\mathbf{C})$ defines a **pullback** γ'^* on homogeneous polynomials (i.e., on 0-forms) by composition: $\gamma'^*(F) = F \circ \gamma'$, cf. Section 4.4. Extend this pullback operator from 0-forms to general forms by having it commute with the differentiation operator d. In particular, for any $\gamma' \in \mathrm{GL}_2(\mathbf{C})$, letting $dZ = \begin{bmatrix} dZ_1 \\ dZ_2 \end{bmatrix}$ denote componentwise differentiation of the vector $Z = \begin{bmatrix} Z_1 \\ Z_2 \end{bmatrix}$ gives

$$\gamma'^*(dZ) = d(\gamma'^* Z) = d(Z \circ \gamma') = d(\gamma' Z) = \gamma' dZ, \tag{7.5.2}$$

with the last equality holding because differentiation is a linear operator on 0-forms. (Exercise 7.5.3 asks for this string of equalities with the matrices and vectors written out, which may clarify what is going on.) More generally, any 1-form is an inner product

$$\mu = F \cdot dZ$$

where $F = \begin{bmatrix} F_1 \\ F_2 \end{bmatrix}$ and again $dZ = \begin{bmatrix} dZ_1 \\ dZ_2 \end{bmatrix}$. (The inner product is denoted "$\cdot$" for brevity, rather than $\langle , \rangle$ as in Chapter 2.) Using this notation and citing—or defining—that the pullback of a product is the product of the pullbacks, the pullback of any $\gamma' \in \mathrm{GL}_2(\mathbf{C})$ on a 1-form is $\gamma'^* \mu = (F \circ \gamma') \cdot \gamma' dZ$, or, thanks to the relation between transpose and inner product explained in Section 2.2,

$$\gamma'^*\mu = \gamma'^t(F \circ \gamma') \cdot dZ. \tag{7.5.3}$$

For example, this formula shows that the 1-form $\lambda = -Z_2 dZ_1 + Z_1 dZ_2$ pulls back under $\gamma' \in \mathrm{PGL}_2(\mathbf{C})$ to $\det(\gamma')\lambda$ (Exercise 7.5.4). In particular, $\gamma'^*\lambda = \lambda$ for all $\gamma' \in \mathrm{SL}_2(\mathbf{C})$. This 1-form λ will play an important role in the next section. As for 2-forms, if $\gamma' = \begin{bmatrix} a & b \\ c & \delta \end{bmatrix} \in \mathrm{SL}_2(\mathbf{C})$ then by various results,

$$\gamma'^*(dZ_1 dZ_2) = d(aZ_1 + bZ_2)d(cZ_1 + \delta Z_2) = (a\delta - bc)dZ_1 dZ_2 = dZ_1 dZ_2,$$

and in general $\gamma'^*(F dZ_1 dZ_2) = (F \circ \gamma')dZ_1 dZ_2$. Since the 1-form λ and the basic 2-form $dZ_1 dZ_2$ are so nicely invariant under pullback by matrices of determinant 1, we will renormalize and work in $\mathrm{PSL}_2(\mathbf{C})$ and $\mathrm{SL}_2(\mathbf{C})$ in the next section to exploit the invariance.

These few results are all that we need about forms, but of course there is much more to be said on the subject. Forms are ubiquitous in areas of mathematics such as differential geometry and differential topology, for example. To get started on the general theory, see Spivak [Sp] or Chapter 10 of Rudin [Ru].

Exercises

7.5.1. Use the skew symmetry rule to show that $dZ_1 dZ_1 = dZ_2 dZ_2 = 0$ and $dZ_2 dZ_1 = -dZ_1 dZ_2$. Use these identities and other algebraic rules for multiplication as necessary to verify equations (7.5.1).

7.5.2. Show that $d^2 F = 0$ for any 0-form F. Thus if $\mu = dF$ then $d\mu = 0$. Conversely, show that if $d\mu = 0$ for a 1-form μ then $\mu = dF$ for some 0-form F.

7.5.3. Letting $\gamma' = \begin{bmatrix} a & b \\ c & \delta \end{bmatrix}$ and $dZ = \begin{bmatrix} dZ_1 \\ dZ_2 \end{bmatrix}$, write out equations (7.5.2) in coordinates, justifying each step. (For the third equality, remember that Z_1 and Z_2 are component functions and γ' represents a vector of linear maps.)

7.5.4. Let $\lambda = -Z_2 dZ_1 + Z_1 dZ_2 = F \cdot dZ$ where $F = \begin{bmatrix} -Z_2 \\ Z_1 \end{bmatrix}$. Show that for any $\gamma \in \mathrm{GL}_2(\mathbf{C})$, $\gamma'^*\lambda = \det(\gamma')\lambda$. (By (7.5.3), it suffices to show that $\gamma'^t(F \circ \gamma') = \det(\gamma')F$. Let $\gamma' = \begin{bmatrix} a & b \\ c & \delta \end{bmatrix}$ and compute. See the hint for the preceding exercise if necessary.)

6. Normal rational functions

Given any $\gamma \in \mathrm{PSL}_2(\mathbf{C})$, this section shows how to compute all rational functions f that commute with γ. Call such functions **γ-normal.** If Γ is a subgroup of $\mathrm{PSL}_2(\mathbf{C})$ and f is γ-normal for each $\gamma \in \Gamma$ then f is **Γ-normal.**

To use differential forms to find all Γ-normal rational functions, identify each rational function $f = f_1/f_2$ with its **dual** 1-form as follows: let F_i be the homogenization of f_i to degree $\max\{\deg(f_1), \deg(f_2)\}$ for $i = 1, 2$; let $F = \begin{bmatrix} F_1 \\ F_2 \end{bmatrix}$ and $dZ = \begin{bmatrix} dZ_1 \\ dZ_2 \end{bmatrix}$ as usual, and let $S = \begin{bmatrix} 0 & -1 \\ 1 & 0 \end{bmatrix}$ be the **skew matrix**, with the property that $Sm^{-1} = m^t S$ for all $m \in \mathrm{SL}_2(\mathbf{C})$ (Exercise 7.6.1). Then the dual 1-form to f is

$$\mu = -F_2 dZ_1 + F_1 dZ_2 = SF \cdot dZ.$$

Note that since f_1 and f_2 are only defined up to a nonzero scalar multiple, μ is also defined only up to scalar, and conversely, all scalar multiples of μ represent the same rational function f. For any $\gamma \in \mathrm{PSL}_2(\mathbf{C})$ with lift $\gamma' \in \mathrm{SL}_2(\mathbf{C})$, the composition $\gamma^{-1} f \gamma$ is identified with the form

$$\begin{aligned} S(\gamma'^{-1} F \circ \gamma') \cdot dZ &= S\gamma'^{-1}(F \circ \gamma') \cdot dZ \\ &= \gamma'^t (SF \circ \gamma') \cdot dZ \qquad && \text{since } S\gamma'^{-1} = \gamma'^t S \\ &= \gamma'^* \mu && \text{by (7.5.3).} \end{aligned}$$

In other words, conjugating f by γ corresponds to pulling back its dual 1-form μ by γ'. This handy fact reduces finding γ-normal rational functions f to finding 1-forms μ such that $\gamma'^* \mu = \chi(\gamma')\mu$ for some nonzero $\chi(\gamma') \in \mathbf{C}$. The next proposition does so.

(7.6.1) PROPOSITION. *Let μ be a 1-form of degree n and let $\gamma' \in \mathrm{SL}_2(\mathbf{C})$. Then*

$$\gamma'^* \mu = \chi(\gamma')\mu \quad \textit{for some } \chi(\gamma') \in \mathbf{C}^*$$

if and only if

$$\mu = G\lambda + dH,$$

where $G \in \mathbf{C}_{n-1}[Z_1 : Z_2]$ and $H \in \mathbf{C}_{n+1}[Z_1 : Z_2]$ satisfy $\gamma'^* G = \chi(\gamma')G$ and $\gamma'^* H = \chi(\gamma')H$, and $\lambda = -Z_2 dZ_1 + Z_1 dZ_2$ is the $\mathrm{SL}_2(\mathbf{C})$-invariant 1-form from Section 7.5.

PROOF. If μ has the described form then indeed μ is a 1-form of degree n, and

$$\begin{aligned}\gamma'^{*}\mu &= (\gamma'^{*}G)(\gamma'^{*}\lambda) + \gamma'^{*}(dH) = (\chi(\gamma')G)\lambda + d(\gamma'^{*}H)\\ &= \chi(\gamma')(G\lambda) + d(\chi(\gamma')H) = \chi(\gamma')(G\lambda + dH) = \chi(\gamma')\mu.\end{aligned}$$

Conversely, suppose $\gamma'^{*}\mu = \chi(\gamma')\mu$. The derivative $d\mu$ takes the form KdZ_1dZ_2 for some 0-form K; let $k = \deg(K)$, and let $G = K/(k+2)$. Compute that

$$\begin{aligned}d(G\lambda) &= d(-Z_2GdZ_1 + Z_1GdZ_2) = (Z_1D_1G + Z_2D_2G + 2G)dZ_1dZ_2\\ &= (k+2)GdZ_1dZ_2 = d\mu,\end{aligned}$$

with the third equality due to Euler's identity from Exercise 1.4.5. (Warning: the "d" in the exercise is the degree k here, not the differentiation operator.) It follows that $d(\mu - G\lambda) = 0$, showing that $\mu = G\lambda + dH$ for some 0-form H. The relations among $\deg(G)$, $\deg(H)$ and $\deg(\mu)$ follow from the nature of differentiation. Finally, since differentiation commutes with the pullback and dZ_1dZ_2 is γ'-invariant,

$$\begin{aligned}(\gamma'^{*}K)dZ_1dZ_2 &= \gamma'^{*}(KdZ_1dZ_2) = \gamma'^{*}d\mu = d\gamma'^{*}\mu = d(\chi(\gamma')\mu) = \chi(\gamma')d\mu\\ &= \chi(\gamma')KdZ_1dZ_2;\end{aligned}$$

So $G = K/(k+2)$ transforms by $\chi(\gamma')$ under γ'^{*}. Since $dH = \mu - G\lambda$, a similar argument (Exercise 7.6.2) shows that H also transforms by $\chi(\gamma')$. □

Let Γ be a subgroup of $\mathrm{PSL}_2(\mathbf{C})$ with lift $\Gamma' \subset \mathrm{SL}_2(\mathbf{C})$. The procedure for producing Γ-normal rational functions is now clear from Proposition 7.6.1: find Γ'-invariant forms G and H that transform by the same character $\chi : \Gamma' \longrightarrow \mathbf{C}^*$ and with $\deg(H) = \deg(G) + 2$; write the form $G\lambda + dH$ as $-F_2dZ_1 + F_1dZ_2$; then the dehomogenized quotient $f = (F_1)_*/(F_2)_*$ is Γ-normal. Since the proposition is bidirectional, this process yields all Γ-normal rational functions. When $dH = 0$, the resulting rational function is the identity, $f = Z$ (Exercise 7.6.3).

See [Do-Mc] for an elegant geometric construction of functions that commute with the icosahedral group. Another construction from [Do-Mc], using a variant of Newton's method, will be discussed below.

Exercises

7.6.1. Show that $Sm^{-1} = m^tS$ for all $m \in \mathrm{SL}_2(\mathbf{C})$.

7.6.2. Confirm the details of the proof of Proposition 7.6.1 as necessary. In particular, show that H transforms under γ'^{*} by $\chi(\gamma')$.

7.6.3. Show that the process described after Proposition 7.6.1 yields the identity function when $dH = 0$.

7. Ingredients of the algorithm

Recall the problem we are working on: our given is the symbol W', which parametrizes the Brioschi quintic

$$b = b_{W'} = T^5 - 10W'T^3 + 45W'^2T - W'^2,$$

a polynomial over the coefficient field $\mathbf{C}(W')$. We want a generally convergent W'-parametrized iteration $F = F_{W'}$ over $\mathbf{C}(W')$ from whose output we can compute a Brioschi root. The Brioschi splitting field extension is $\mathbf{C}(Z)/\mathbf{C}(W')$, where the splitting field generator Z and the Brioschi parameter W' are related by the icosahedral invariant function and a fractional linear transformation. Specifically, the icosahedral invariant is the degree-60 rational function $f_I(Z)$ constructed in Section 3.6, also denoted W; and the fractional linear relation between W and W', from Section 4.8, is $W' = 1/(1728(1 - W))$. The icosahedral invariant W and the Brioschi parameter W' are essentially the same: transforming between them merely renormalizes the invariant, which was noncanonical in the first place, and since the transformation is field-theoretically trivial, the Brioschi splitting field extension is also $\mathbf{C}(Z)/\mathbf{C}(W)$. This is the icosahedral extension; that is, Theorem 3.6.2 says that $\mathbf{C}(W)$ is the subfield of $\mathbf{C}(Z)$ invariant under the icosahedral group Γ_I, and Section 4.4 shows further that $\mathrm{Gal}(\mathbf{C}(Z)/\mathbf{C}(W))$ is precisely Γ_I.

To find an iteration $F \in \mathbf{C}(W')(T)$ that gives a Brioschi root, the following data are necessary according to the second half of Theorem 7.3.5: simply taking the embedding $\rho : \Gamma_I \longrightarrow \mathrm{PGL}_2(\mathbf{C})$ to be the identity map, we need (1) a generally convergent Γ_I-normal model $f \in \mathbf{C}(T)$, and (2) a conjugating transformation $\phi \in \mathrm{PGL}_2(\mathbf{C}(Z))$ such that $\phi^\gamma \phi^{-1} = \gamma$ for all $\gamma \in \Gamma_I$. This conjugating transformation ϕ is a rationally Z-parametrized fractional linear transformation of the iteration variable T. In more expanded notation, ϕ is

$$\phi = \phi_Z = \begin{bmatrix} a(Z) & b(Z) \\ c(Z) & \delta(Z) \end{bmatrix} \in \mathrm{PGL}_2(\mathbf{C}(Z))$$

as a transformation, that is,

$$\phi_Z(T) = \frac{a(Z)T + b(Z)}{c(Z)T + \delta(Z)} \in \mathbf{C}(Z)(T).$$

The required relation $\phi^\gamma \phi^{-1} = \gamma$ rewrites as

$$\gamma \circ \phi_Z = \phi_Z^\gamma = \phi_{\gamma Z} \quad \text{for all } \gamma \in \Gamma_I.$$

In other words, ϕ is a Z-parametrized family of fractional linear transformations whose dependence on their parameter Z is Γ_I-normal. We will notate ϕ_Z with a subscript when we care how it depends on its parameter Z; when we care primarily how ϕ acts on the iteration variable T, the subscript will be omitted.

This section computes a rational function f and a Z-parametrized conjugating transformation ϕ with the right transformation properties. The next section shows that in fact the second iterate $f \circ f$, rather than f itself, has the requisite convergence properties to serve as a model, so that our algorithm is to iterate the composite $F = \phi^{-1} f \phi$ an even number of times. Section 7.9 carries out one last calculation to parametrize the composite $F = \phi_Z^{-1} f \phi_Z$ by the Brioschi parameter W' rather than by the parameter Z of the conjugating transformation. This is crucial because W' is our given datum while Z, the Brioschi splitting field generator, is certainly not given—if it were, we would have all the Brioschi roots, rendering the entire problem null; thus, we have not found F in any useful sense until it is expressed in terms of W'. Finally, Section 7.10 computes a Brioschi root (in fact, two Brioschi roots) from the output of the iteration F.

Since the model f and the conjugating transformation ϕ must both be Γ_I-normal, the methods of the preceding section construct them. The Γ_I-invariant forms were computed in Chapter 3; they are generated by the vertex form $F_{1,I}$ of degree 12, the face-center form $F_{2,I}$ of degree 20, and the mid-edge form $F_{3,I}$ of degree 30. No two of these have degrees differing by 2, so the only way to produce nonidentity Γ_I-normal rational functions from the generators is to set $G = 0$, which may be viewed as having any desired degree and transforming by any desired character, and $H = F_{1,I}$ or $H = F_{2,I}$ or $H = F_{3,I}$. (Setting $H = 0$ produces the identity function $f = Z$, by Exercise 7.6.3.) The case $H = F_{1,I}$ is the simplest, giving a rational function of degree 11 (Exercise 7.7.1),

$$f_{11} = -\frac{(D_2 F_{1,I})_*}{(D_1 F_{1,I})_*} = -\frac{(Z_1^{11} + 66 Z_1^6 Z_2^5 - 11 Z_1 Z_2^{10})_*}{(11 Z_1^{10} + 66 Z_1^5 Z_2^6 - Z_2^{10})_*} = -\frac{T^{11} + 66T^6 - 11T}{11T^{10} + 66T^5 - 1}.$$

Here $[Z_1 : Z_2]$ is dehomogenized to T rather than to Z because this f_{11}—or rather, its second iterate $f_{11} \circ f_{11}$, as to be discussed in the next section—will serve as the model for the iterative solution of the Brioschi quintic, and T is the iteration variable.

The procedure that produced f_{11} also produces ϕ with a bit more work. Note that the icosahedral forms $F_{3,I}$ and $F_{1,I}F_{2,I}$ have respective degrees 30 and 32. Therefore a one-parameter family of forms,

$$T_1 F_{3,I}\lambda + T_2 d(F_{1,I}F_{2,I}) \quad \text{as } [T_1 : T_2] \text{ varies through } \mathbf{P}^1(\mathbf{C}),$$

is Γ_I-invariant, giving rise to a corresponding family of rational functions. These work out to

$$f_T(Z) = \frac{(T_1 Z_1 F_{3,I} + T_2 D_2(F_{1,I}F_{2,I}))_*}{(T_1 Z_2 F_{3,I} - T_2 D_1(F_{1,I}F_{2,I}))_*},$$

where the dehomogenization takes the variable $[Z_1 : Z_2]$ to Z and the parameter $[T_1 : T_2]$ to T (Exercise 7.7.3). Each T-parametrized rational function f_T is Γ_I-normal as a function of its variable Z, by construction. Now comes the cunning idea: we reverse our notions of what is a variable and what is a parameter, viewing f as a Z-parametrized fractional linear transformation acting on the iteration variable T,

(7.7.1)

$$\begin{aligned} f_T(Z) &= \frac{(Z_1 F_{3,I} T_1 + D_2(F_{1,I}F_{2,I})T_2)_*}{(Z_2 F_{3,IT_1} - D_1(F_{1,I}F_{2,I})T_2)_*} \\ &= \begin{bmatrix} Z_1 F_{3,I} & D_2(F_{1,I}F_{2,I}) \\ Z_2 F_{3,I} & -D_1(F_{1,I}F_{2,I}) \end{bmatrix}_* (T) \overset{\text{call}}{=} \psi_Z(T). \end{aligned}$$

In this new notation, the fact that f_T is Γ_I-normal rewrites as $\gamma \circ \psi_Z = \psi_{\gamma Z} = \psi_Z^{\gamma}$ for all $\gamma \in \Gamma_I$, which is precisely how we need the conjugating transformation to behave. More generally, the same relation holds for $\psi \circ m$ where $m \in \mathrm{PGL}_2(\mathbf{C})$ is any fixed fractional linear transformation; that is (Exercise 7.7.4),

$$\gamma \circ (\psi_Z \circ m) = (\psi_Z \circ m)^{\gamma} \quad \text{for all } \gamma \in \Gamma_I.$$

So any $\psi \circ m$ is a candidate for the conjugating transformation of an iterative algorithm to solve the Brioschi quintic.

Next we normalize m to obtain a particularly convenient conjugating transformation $\phi = \psi \circ m$. Applying the product rule to the second column of ψ in (7.7.1) shows that

$$\begin{aligned} \psi &= \left[F_{3,I} \begin{bmatrix} Z_1 \\ Z_2 \end{bmatrix} \quad F_{2,I} \begin{bmatrix} D_2 F_{1,I} \\ -D_1 F_{1,I} \end{bmatrix} + F_{1,I} \begin{bmatrix} D_2 F_{2,I} \\ -D_1 F_{2,I} \end{bmatrix} \right]_* \\ &= \begin{bmatrix} c_1 & c_2 + c_2' \end{bmatrix}_* \end{aligned}$$

where c_1, c_2, c_2' are column vectors. It turns out that c_2' is a linear combination of c_1 and c_2:

(7.7.2) LEMMA. $c_2' = \frac{5}{3}(c_1 + c_2)$.

The proof is Exercise 7.7.5. Now let $m = \begin{bmatrix} 1 & -5/8 \\ 0 & 3/8 \end{bmatrix}$. Right-multiplication by m carries out the column operations to make $\phi = \psi \circ m$ simply $\phi = \begin{bmatrix} c_1 & c_2 \end{bmatrix}_*$ (Exercise 7.7.6), i.e.,

$$\phi_Z = \begin{bmatrix} Z_1 F_{3,I} & F_{2,I} D_2 F_{1,I} \\ Z_2 F_{3,I} & -F_{2,I} D_1 F_{1,I} \end{bmatrix}_* .$$

Again, the subscript $_*$ means to dehomogenize $[Z_1 : Z_2]$ to Z.

The ingredients for a generally convergent algorithm are all in place: the model will be $f_{11} \circ f_{11}$ and the conjugating transformation will be ϕ. Both commute appropriately with the icosahedral group Γ_I. To apply Theorem 7.3.5, the next section will establish convergence properties of the model.

Exercises

7.7.1. Confirm the computation of the Γ_I-normal rational function f_{11}, and compute the corresponding f_{19} and f_{29}.

7.7.2. Confirm directly that f_{11} commutes with the icosahedral generators s_I and t_I from Section 2.7.

7.7.3. Confirm the formula for $f_T(Z)$.

7.7.4. Let ψ and m be as in the section. Explain why $\gamma \circ (\psi_Z \circ m) = (\psi_Z \circ m)^\gamma$ for all $\gamma \in \Gamma_I$.

7.7.5. To prove Lemma 7.7.2, start from the syzygy $1728F_{1,I}^5 - F_{2,I}^3 - F_{3,I}^2 = 0$. Take the gradient (vector of partial derivatives), multiply through by $F_{1,I}/F_{2,I}^2$, substitute $1728F_{1,I}^5 = F_{2,I}^3 + F_{3,I}^2$, and rearrange to get (where ∇ denotes gradient):

$$5F_{2,I}\nabla F_{1,I} - 3F_{1,I}\nabla F_{2,I} = F_{3,I}[2(F_{1,I}/F_{2,I}^2)\nabla F_{3,I} - 5(F_{3,I}/F_{2,I}^2)\nabla F_1].$$

Since the left side is a vector of degree-31 forms, the quantity in brackets on the right must be a vector of degree-1 forms. Compute the highest Z_1-order and Z_2-order terms in the left side to find the bracketed quantity and complete the proof.

7.7.6. Confirm that $\psi \circ m = \phi$ as claimed in the section.

7.7.7. Confirm directly that $\phi_Z(T)$ commutes with the icosahedral generators s_I and t_I from Section 2.7 as a function of Z.

8. General convergence of the model

The task at hand now is to discuss convergence properites of f_{11}. Recall that f_{11} is the dehomogenization of the quotient $-D_2H/D_1H$ where $H = F_{1,I}$ is the icosahedral vertex form. Thanks to Euler's identity, this is

$$f_{11} = \left(\frac{Z_1D_1H - Z_1D_1H - Z_2D_2H}{Z_2D_1H}\right)_* = \left(\frac{Z_1}{Z_2} - \frac{\deg(H)H}{Z_2D_1H}\right)_* = T - 11\frac{h}{h'}$$

where $h = H_*$. (Exercise 7.8.1 asks for a direct verification of this.) Thus f_{11} is computed by a variant of Newton's method on the dehomogenization h. In particular, the fixed points of f_{11} are the roots of h, i.e., the icosahedral vertices. At those fixed points, $|f_{11}'|$ works out to $\deg(h) - 1 = 10$ (Exercise 7.8.2); since this is greater than 1, iteration theory unfortunately says that the fixed points of f_{11} are **repelling**, i.e., nearby points iterate away from them. Since the attractor of f_{11} must consist of fixed points, we have just shown it is empty, and f_{11} is highly nonconvergent.

To address this, broaden the notion of fixed point by defining a **cycle** of a rational function $f \in \mathbf{C}(T)$ to be a set $\{a_1, a_2, \ldots, a_n\} \subset \widehat{\mathbf{C}}$ such that

$$f(a_1) = a_2, \quad f(a_2) = a_3, \quad \ldots, \quad f(a_{n-1}) = a_n, \quad f(a_n) = a_1.$$

In the case of f_{11}, each opposing pair of icosahedral face-centers forms a cycle. To see this, let a be a face-center and let $\Gamma_a \subset \Gamma_I$ be its stabilizer, of order 3. Then $\Gamma_a(f_{11}(a)) = f_{11}(\Gamma_a(a)) = f_{11}(a)$, so Γ_a also stabilizes $f_{11}(a)$. Therefore $f_{11}(a)$, being distinct from a, must be its antipode, the only other Γ_a-stable point.

The same argument shows that the fifteen opposing pairs of icosahedral mid-edge points form f_{11}-cycles (Exercise 7.8.3). Consequently the function $f_{11} \circ f_{11}$ fixes all the icosahedral vertices, face-centers and mid-edges. These three families of fixed points have significantly different properties, however; to investigate the difference, define a **critical point** of any rational function $f \in \mathbf{C}(T)$ to be a point $a \in \widehat{\mathbf{C}}$ such that $f'(a) = 0$.

(7.8.1) LEMMA. *Suppose the rational function f takes the form $f = [G : H]$ as an algebraic mapping, cf. Section 1.4. Then*

$$f' = 0 \quad \Longleftrightarrow \quad \det\begin{bmatrix} D_1G & D_2G \\ D_1H & D_2H \end{bmatrix} = 0.$$

The proof is Exercise 7.8.4. In particular, this shows that the critical points of $f_{11} = [-D_2F_{1,I} : D_1F_{1,I}]$ are precisely the roots of the Hessian of $F_{1,I}$ from Section 3.3 (Exercise 7.8.5). This Hessian is the icosahedral face-center form $F_{2,I}$, so the critical points of f_{11} are the twenty icosahedral face-centers. (Exercise 7.8.6 shows this in another way.) The vertices and mid-edges, on the other hand, are not critical points.

Thus the twenty critical points of f_{11} form ten 2-cycles of f_{11}. This sets up the result that $f_{11} \circ f_{11}$ is generally convergent. Define a cycle of any rational function f to be **superattracting** if it contains a critical point of f, and define f itself to be **critically finite** if each critical point t is taken to a cycle by finitely many iterations of f. In particular, f_{11} is critically finite since every critical point is already in a cycle. The result from dynamics that we need is

(7.8.2) THEOREM. *Let f be critically finite and let A be the union of its superattracting cycles. If A is nonempty then for all initial guesses t in some dense open subset of $\widehat{\mathbf{C}}$, the iteration $\{f^k(t)\}$ converges to a cycle in A.*

See Theorem 3.3 of [Do-Mc] for a stronger statement and the relevant references.

Since f_{11} is critically finite and all of its superattracting cycles have length 2, Theorem 7.8.2 shows that $f_{11} \circ f_{11}$ is generally convergent to the icosahedral face-centers. The theorem also shows that Newton's method for roots of unity is generally convergent (Exercise 7.8.8).

The icosahedral group Γ_I acts transitively on the set $A = \text{Att}(f_{11} \circ f_{11})$, as this set comprises the icosahedral face-centers; so all hypotheses for the second half of Theorem 7.3.5 are now satisfied. The theorem says that for general values of the Brioschi parameter w', iterating the algorithm

$$F_{w'}(T) = (\phi_z^{-1} f_{11} \phi_z)(T) \in \mathbf{C}(T) \qquad (\text{where } w' = 1/(1728(1 - f_I(z))))$$

an even number of times—or, for that matter, an odd number of times—converges for general values of the starting guess t. According to the end of the theorem, the resulting output field constructed over $\mathbf{C}(W')$ is $\mathbf{C}(\phi_Z^{-1}(a))$ for any icosahedral face-center a, the fixed field in $\mathbf{C}(Z)$ of the corresponding face-center stabilizing subgroup $\Gamma_a \subset \Gamma_I$. As we will see in Section 7.10, this field contains two Brioschi roots.

Exercises

7.8.1. Verify directly that $f_{11} = T - 11h/h'$ where h is the dehomogenization of $F_{1,I}$.

7.8.2. Show that $f'_{11} = 1 - \deg(h) = -10$ at the fixed points.

7.8.3. Show that the fifteen opposing pairs of icosahedral mid-edge points form f_{11}-cycles.

7.8.4. Prove Lemma 7.8.1 as follows: the numerator of f' is $(HD_1G - GD_1H)_*$, which by Euler's identity is a scalar multiple of $(D_1GD_2H - D_2GD_1H)_*$; this is the determinant, dehomogenized.

7.8.5. Use Lemma 7.8.1 to show that the critical points of f_{11} are precisely the roots of the Hessian of $F_{1,I}$.

7.8.6. Give another proof that the critical points of f_{11} are the icosahedral face-centers as follows. For any subgroup $\Gamma \in \mathrm{PSL}_2(\mathbf{C})$ and any Γ-normal f, the critical points of f form a Γ-invariant set. The equation $f'_{11} = 0$ has degree 20, so the result follows geometrically.

7.8.7. Use a computer to iterate f_{11} experimentally on random inputs and observe its general convergence to 2-cycles.

7.8.8. Use Theorem 7.8.2 to show that Newton's method for nth roots of unity (see Section 7.1) is generally convergent to nth roots of unity.

7.8.9. As explained at the end of the section, the output field constructed by F over $\mathbf{C}(W')$ is $\mathbf{C}(\phi_Z^{-1}(a))$ for any icosahedral face-center a, the fixed field in $\mathbf{C}(Z)$ of the face-center stabilizing subgroup $\Gamma_a \subset \Gamma_I$. What are the extension degrees $[\mathbf{C}(\phi_Z^{-1}(a)) : \mathbf{C}(W')]$ and $[\mathbf{C}(Z) : \mathbf{C}(\phi_Z^{-1}(a))]$?

9. Computing the algorithm

Possessing explicit formulas for the model f_{11} and the conjugating transformation ϕ, we consequently possess an explicit formula for the composite algorithm $F = \phi^{-1} f_{11} \phi$. One problem remains, as explained in Section 7.7: though the theory shows that F is a W'-parametrized rational function of the iteration variable T, where W' is the given Brioschi coefficient field generator, our current expression for F is Z-parametrized, where Z is the unknown Brioschi splitting field generator. We need to express F in terms of W' for it to be of any use.

Recall the derivation of f_{11},

$$\begin{aligned} f_{11} &= (-D_2H)_*/(D_1H)_* && \text{where } H = F_{1,I}, \text{ the icosahedral vertex-form} \\ &= T - \deg(h)h/h' && \text{where } h = H_*, \text{ dehomogenizing } (Z_1, Z_2) \text{ to } T. \end{aligned}$$

The second formula shows that f_{11} fixes each point where H vanishes, i.e., the set A of icosahedral vertices.

We now obtain similar formulas for F. Lift ϕ and homogenize in Z and

T to obtain a (Z_1, Z_2)-parametrized linear function of (T_1, T_2),

$$\Phi_{(Z_1,Z_2)}(T_1,T_2) = \begin{bmatrix} Z_1F_{3,I} & F_{2,I}D_2F_{1,I} \\ Z_2F_{3,I} & -F_{2,I}D_1F_{1,I} \end{bmatrix} \begin{bmatrix} T_1 \\ T_2 \end{bmatrix}.$$

Redefine H to be a (Z_1, Z_2)-parametrized function of (T_1, T_2),

$$H_{(Z_1,Z_2)}(T_1,T_2) = \frac{(F_{1,I} \circ \Phi_{(Z_1,Z_2)})(T_1,T_2)}{F_{1,I}(Z_1,Z_2)F_{3,I}(Z_1,Z_2)^{12}}.$$

This warrants some motivation. The numerator of H is a parametrically transformed version of the previous $H = F_{1,I}$, designed to vanish on the set $\Phi^{-1}_{(Z_1,Z_2)}(A)$ (where the set A of icosahedral vertices has been suitably lifted from $\mathbf{P}^1(\mathbf{C})$ to $\mathbf{C}^2$), precisely the set that F needs to fix. Also, the numerator is Γ'_I-invariant in its parameter (Z_1, Z_2) because Φ is Γ'_I-normal and $F_{1,I}$ is Γ'_I-invariant, so for any $\gamma' \in \Gamma'_I$,

$$F_{1,I} \circ \Phi_{\gamma'(Z_1,Z_2)} = F_{1,I} \circ \gamma' \circ \Phi_{(Z_1,Z_2)} = F_{1,I} \circ \Phi_{(Z_1,Z_2)}.$$

Clearly the vanishing property also holds for the quotient H, and so does the invariance property since the denominator is Γ'_I-invariant. This doesn't yet explain the strange denominator, but regardless of its value, the vanishing and invariance properties of H suggest—and a calculation confirms (Exercise 7.9.1)—that $F = \phi^{-1}f_{11}\phi$ may be computed by the same method as f_{11}. Specifically,

(7.9.1)

$$F = (-D_2H)_*/(D_1H)_* = T - \deg(h)h/h' \qquad \text{where } h = H_*,$$

where the partial derivatives are with respect to the variables T_1, T_2, which dehomogenize to the iteration variable T while Z_1, Z_2 dehomogenize to the parameter Z. Therefore, parametrizing F by the Brioschi parameter W' rather than Z reduces to doing the same for h/h'. The denominator of H, which is simply a constant as far as T_1, T_2 are concerned and therefore cancels in the quotient h/h' anyway, has been concocted to make h rather than h/h' parametrized by W', as we will see soon. This will simplify the notation once we compute h.

Renormalize the icosahedral invariant Brioschi parameter W' by a fractional linear transformation to $\widehat{W} = 1 - 1728W'$. The defining relations $W' = 1/(1728(1-W))$, $W = f_I(Z) = (F^3_{2,I})_*/(1728F^5_{1,I})_*$ and the syzygy $1728F^5_{1,I} - F^3_{2,I} - F^2_{3,I} = 0$ show that $\widehat{W} = (-F^3_{2,I})_*/(F^2_{3,I})_*$ (Exercise 7.9.2). No icosahedral invariant is canonical—we have already discussed this point in connection with the Brioschi parameter W' and the original icosahedral invariant $W = f_I(Z)$—and $\widehat{W}$ makes the calculations cleaner, so we will use it instead of W' from now on. In particular, the iteration is now renamed $F_{\widehat{W}}$.

To express $h = H_*$ in terms of $\widehat{W}$, first note that the numerator of H, which is the composition $F_{1,I} \circ \Phi$, has degree $12 \cdot 31$ in (Z_1, Z_2) and degree 12 in (T_1, T_2). Since $12 \cdot 31 \equiv 12 \pmod{60}$ and the Γ_I'-invariant forms are generated by $-F_{2,I}$ and $F_{3,I}$, arguing by degree shows that the numerator takes the form

$$F_{1,I} \circ \Phi = F_{1,I} \sum_{i=0}^{6} \sum_{j=0}^{12} a_{ij} (-F_{2,I})^{3i} F_{3,I}^{12-2i} T_1^j T_2^{12-j}, \tag{7.9.2}$$

where the 91 coefficients a_{ij} can be found by solving a formidable system of 4849 linear equations. Equation (7.9.2) rewrites as

$$F_{1,I} \circ \Phi = F_{1,I} F_{3,I}^{12} T_2^{12} \sum_{i=0}^{6} \sum_{j=0}^{12} a_{ij} \left(\frac{-F_{2,I}^3}{F_{3,I}^2} \right)^i \left(\frac{T_1}{T_2} \right)^j,$$

and since $(T_2)_* = 1$, this relation dehomogenizes to

$$\left(\frac{F_{1,I} \circ \Phi}{F_{1,I} F_{3,I}^{12}} \right)_* = \sum_{i=0}^{6} \sum_{j=0}^{12} a_{ij} \widehat{W}^i T^j \overset{\text{call}}{=} h_{\widehat{W}}(T).$$

This is precisely H_*, explaining where the denominator came from when we defined H in the first place: without it, h would not be parametrized by $\hat{w}$. Specifically, h works out to (Exercise 7.9.3)

$$\begin{aligned} h_{\widehat{W}} = {} & 91125\widehat{W}^6 + (-133650T^2 + 61560T - 193536)\widehat{W}^5 \\ & + (-66825T^4 + 142560T^3 + 133056T^2 - 61440T + 102400)\widehat{W}^4 \\ & + (5940T^6 + 4752T^5 + 63360T^4 - 140800T^3)\widehat{W}^3 \\ & + (-1485T^8 + 3168T^7 - 10560T^6)\widehat{W}^2 \\ & + (-66T^{10} + 440T^9)\widehat{W} + T^{12}. \end{aligned}$$

(Note how many of the a_{ij} are zero.) So we have the iteration parametrized by the icosahedral invariant $\widehat{W}$ as desired,

$$F_{\widehat{W}}(T) = T - 12h_{\widehat{W}}(T)/h'_{\widehat{W}}(T).$$

Exercises

7.9.1. Holding the parameter (Z_1, Z_2) constant (it plays no role in this exercise), write the function $H(T_1, T_2)$ from the section as $H = c \cdot (F_{1,I} \circ \Phi)$

for some constant c. Compute, using the chain rule, that

$$\begin{bmatrix} -D_2H \\ D_1H \end{bmatrix} = c \begin{bmatrix} D_2\Phi_2 & -D_2\Phi_1 \\ -D_1\Phi_2 & D_1\Phi_1 \end{bmatrix} \begin{bmatrix} -D_2F_{1,I} \\ D_1F_{1,I} \end{bmatrix} \circ \Phi.$$

Since Φ is linear, the matrix in this equation is Φ^{-1} up to constant multiple. Dehomogenize and divide to show that $(-D_2H)_*/(D_1H)_* = \phi^{-1}f_{11}\phi$, giving the first equality in (7.9.1); the second equality follows as in the beginning of Section 7.8.

7.9.2. Check that $\widehat{W} = (-F_{2,I}^3)_*/(F_{3,I}^2)_*$.

7.9.3. Write the system of linear equations defining the coefficients a_{ij} of h and either solve them or at least verify h from the section.

10. Solving the Brioschi quintic by iteration

Finally, to proceed from the iteration $F_{\widehat{W}}$ to a root of the Brioschi quintic, we begin by reviewing some of the machinery from Sections 4.8 and 4.9, where the Brioschi resolvent was developed and used to invert the icosahedral equation. (You may want to reread those sections before proceeding here.) Recall that the icosahedral group Γ_I contains a tetrahedral subgroup $\widetilde{\Gamma}_T$ that is a rotated version of the normalized tetrahedral group Γ_T from Chapter 2. Each invariant form for the corresponding rotated octahedral group provides a quintic icosahedral resolvent. The simplest such form is $\tilde{F}_{1,O}$; this form, suitably rationalized to $\tilde{s} = (\tilde{F}_{1,O}F_{1,I}^2/F_{3,I})_* \in \mathbf{C}(Z)$, satisfies the Brioschi resolvent

$$R_{\tilde{s}} = T^5 - 10W'T^3 + 45W'^2T - W'^2,$$

where $W' = 1/(1728(1 - f_I(Z))) = (F_{1,I}^5/F_{3,I}^2)_*$ and now $\widehat{W} = 1 - 1728W'$. The five Brioschi roots are $\tilde{s}^\gamma$ as γ runs through representatives of the coset space $\widetilde{\Gamma}_T\backslash\Gamma_I$. Solving the icosahedral equation from the Brioschi root $\tilde{s}$ was done with the help of the renormalized tetrahedral invariant $\tilde{r} = \tilde{s}^2/W' = (\tilde{F}_{1,O}^2/F_{1,I})_*$. Exercise 4.9.4(a) showed that the twenty icosahedral face-centers divide into three orbits under the tetrahedral group $\widetilde{\Gamma}_T$: the 4-orbit $\mathcal{O}$ of tetrahedral vertices, the 4-orbit $\mathcal{O}'$ of countertetrahedral vertices, and a remaining 12-orbit $\mathcal{P}$ (see Figure 4.9.1). From Exercise 4.9.4(b),

$$\tilde{r}(\mathcal{O}) = \frac{11 + 3i\sqrt{15}}{2}, \quad \tilde{r}(\mathcal{O}') = \frac{11 - 3i\sqrt{15}}{2}, \quad \tilde{r}(\mathcal{P}) = 3.$$

Exercise 4.9.4(c) gave an explicit formula for the quotient $F_{2,I}/\tilde{F}_{2,O}$, which is integral, and showed that $\tilde{r} - 3 = (F_{2,I}/\tilde{F}_{2,O})_*/(F_{1,I})_*$. We will use this tetrahedral invariant to solve the Brioschi quintic.

Let A again denote the attractor of $f_{11} \circ f_{11}$, i.e., the set of icosahedral face-centers. For each value z of the Brioschi splitting field generator Z, the conjugating transformation ϕ_z^{-1} maps A to the cross-section of $\mathrm{Out}(F_{\widehat{W}} \circ F_{\widehat{W}})$ over the corresponding value $\hat{w}$ of the renormalized Brioschi parameter,

$$\phi_z^{-1} : A \longrightarrow \mathrm{Att}(F_{\hat{w}} \circ F_{\hat{w}}),$$

cf. Exercise 7.3.4. Iterating $F_{\hat{w}} \circ F_{\hat{w}}$ on various initial t-values converges to values $\phi_z^{-1}(a)$ for various $a \in A$, as depicted in Figure 7.1.1. Thus $\phi_Z^{-1}(a)$ is our general iteration output. Applying $F_{\hat{w}}$ once more to an output $\phi_z^{-1}(a)$ gives another output $\phi_z^{-1}(-a)$ since $f_{11}(a) = -a$. This is all in accord with the end of Theorem 7.3.5, which says that the output field of $F_{\widehat{W}} \circ F_{\widehat{W}}$ over the Brioschi coefficient field $\mathbf{C}(\widehat{W})$ is $\mathbf{C}(\phi_Z^{-1}(a))$ for any $a \in A$, at the level of function field equivalence. The output field is the fixed field of $\mathrm{stab}(a)$ in the splitting field $\mathbf{C}(Z)$.

Fix any face-center $a \in A$. The group $\mathrm{stab}(a)$ sits in two tetrahedral subgroups of the icosahedral group, the rotation groups for the tetrahedron $\mathcal{T}_a$ with vertex a and the tetrahedron $\mathcal{T}_{-a}$ with vertex $-a$. Note that $\mathcal{T}_{-a}$ is not counter to $\mathcal{T}_a$, cf. Figure 2.5.10. Exercise 7.10.1 asks for proofs of the following assertions: each coset $\widetilde{\Gamma}_T\gamma \in \widetilde{\Gamma}_T\backslash\Gamma_I$ takes a into one of the $\widetilde{\Gamma}_T$-orbits $\mathcal{O}$, $\mathcal{O}'$, $\mathcal{P}$; one coset $\widetilde{\Gamma}_T\gamma_0$ takes a into $\mathcal{O}$, another coset $\widetilde{\Gamma}_T\gamma_1$ takes a into $\mathcal{O}'$, and the other three cosets take a into $\mathcal{P}$. With this notation in place, the tetrahedral subgoups containing $\mathrm{stab}(a)$ are explicitly $\Gamma_{T,0} = \gamma_0^{-1}\widetilde{\Gamma}_T\gamma_0$ and $\Gamma_{T,1} = \gamma_1^{-1}\widetilde{\Gamma}_T\gamma_1$. The corresponding intermediate fields of the Brioschi splitting field extension are generated by appropriate translates of the tetrahedral invariant, $f_{T,0} = \tilde{f}_T^{\gamma_0}$ and $f_{T,1} = \tilde{f}_T^{\gamma_1}$. So we have the lattices shown in Figure 7.10.1.

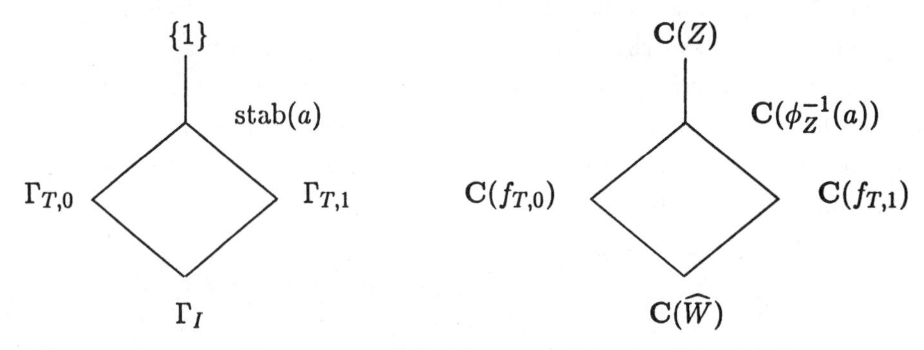

Figure 7.10.1. Groups and fields associated with the iteration

The two Brioschi roots $s_0 = \tilde{s}^{\gamma_0}$ and $s_1 = \tilde{s}^{\gamma_1}$, being translated tetrahedral invariants, lie in the fields $\mathbf{C}(f_{T,0})$ and $\mathbf{C}(f_{T,1})$, so Figure 7.10.1 shows that s_0 and s_1 are rational expressions in $\phi_Z^{-1}(a)$. To find them, exploit the values taken by the tetrahedral invariant $\tilde{r}-3$ on its three orbits by introducing the $\widehat{W}$-parametrized polynomial of T

(7.10.1)

$$\mu_{\widehat{W}}(T) = \sum_{\gamma\in\widetilde{\Gamma}_T\backslash\Gamma_I} \tilde{s}^\gamma(Z)\cdot(\tilde{r}^\gamma\circ\phi_Z(T)-3)\in\mathbf{C}(\widehat{W})[T].$$

(Exercise 7.10.2 asks why μ lies in the subring $\mathbf{C}(\widehat{W})[T]$ of $\mathbf{C}(Z)[T]$.) At our iteration outputs $\phi_Z^{-1}(a)$ and $F_{\widehat{W}}(\phi_Z^{-1}(a)) = \phi_Z^{-1}(-a)$, this polynomial evaluates to (Exercise 7.10.3(a))

$$\mu_{\widehat{W}}(\phi_Z^{-1}(a)) = \frac{5+3i\sqrt{15}}{2}s_0 + \frac{5-3i\sqrt{15}}{2}s_1 \overset{\text{call}}{=} \mu_0,$$
$$\mu_{\widehat{W}}(\phi_Z^{-1}(-a)) = \frac{5-3i\sqrt{15}}{2}s_0 + \frac{5+3i\sqrt{15}}{2}s_1 \overset{\text{call}}{=} \mu_1.$$

Now linear algebra gives the Brioschi roots s_0 and s_1 (Exercise 7.10.3(b)),

$$s_0 = \frac{9-i\sqrt{15}}{90}\mu_0 + \frac{9+i\sqrt{15}}{90}\mu_1, \qquad s_1 = \frac{9+i\sqrt{15}}{90}\mu_0 + \frac{9-i\sqrt{15}}{90}\mu_1.$$

Computing μ explicitly will complete the Brioschi algorithm. To do so, first homogenize, using the formulas for $\tilde{s}$ and $\tilde{r}-3$, and then rationalize,

$$\begin{aligned}
\mu_{\widehat{W}}(T) &= \sum_{\gamma\in\widetilde{\Gamma}_T\backslash\Gamma_I} \tilde{s}^\gamma(Z)\cdot(\tilde{r}^\gamma(\phi_Z(T))-3)\\
&= \left[\sum_{\gamma'\in\widetilde{\Gamma}'_T\backslash\Gamma'_I} \frac{\widetilde{F}^{\gamma'}_{1,O}F^2_{1,I}}{F_{3,I}}\cdot\frac{(F_{2,I}/\widetilde{F}^{\gamma'}_{2,O})\circ\Phi_{(Z_1,Z_2)}(T_1,T_2)}{F_{1,I}\circ\Phi_{(Z_1,Z_2)}(T_1,T_2)}\right]_*\\
&= \left[\frac{\sum_{\gamma'\in\widetilde{\Gamma}'_T\backslash\Gamma'_I}\widetilde{F}^{\gamma'}_{1,O}F_{1,I}\cdot((F_{2,I}/\widetilde{F}^{\gamma'}_{2,O})\circ\Phi_{(Z_1,Z_2)}(T_1,T_2))/F^{13}_{3,I}}{(F_{1,I}\circ\Phi_{(Z_1,Z_2)}(T_1,T_2))/(F_{1,I}F^{12}_{3,I})}\right]_*\\
&= \frac{k_{\widehat{W}}(T)}{h_{\widehat{W}}(T)},
\end{aligned}$$

where h was found in the preceding section and similarly k takes the form (Exercise 7.10.4)

$$k = \sum_{i=0}^{6}\sum_{j=0}^{12} b_{ij}\widehat{W}^iT^j.$$

Solving another large linear system gives (Exercise 7.10.4 again)

$$\begin{aligned} k_{\widehat{W}} = 100\widehat{W}(\widehat{W}-1)\cdot \\ &[(1215T-648)\widehat{W}^4 + (-540T^3 - 216T^2 - 1152T + 640)\widehat{W}^3 \\ &\quad + (378T^5 - 504T^4 + 960T^3)\widehat{W}^2 + (36T^7 - 168T^6)\widehat{W} - T^9]. \end{aligned}$$

This gives the long-awaited purely iterative algorithm for the Brioschi quintic:

(7.10.2) ALGORITHM. *To solve any specific Brioschi quintic* $b_{w'} = T^5 - 10w'T^3 + 45w'^2T - w'^2$ *with* $w' \in \mathbf{C}$,

1. *Let* $\hat{w} = 1 - 1728w'$ *and compute the polynomials* $h_{\hat{w}}, k_{\hat{w}} \in \mathbf{C}[T]$.
2. *Iterate the rational function* $F_{\hat{w}} = T - 12h_{\hat{w}}/h'_{\hat{w}} \in \mathbf{C}(T)$ *an even number of times on a random initial guess* t *until the iteration converges to some value* t_0. *Set* $t_1 = F_{\hat{w}}(t_0)$.
3. *Set* $\mu_0 = k_{\hat{w}}(t_0)/h_{\hat{w}}(t_0)$ *and* $\mu_1 = k_{\hat{w}}(t_1)/h_{\hat{w}}(t_1)$.
4. *Then*

$$s_0 = \frac{9 - i\sqrt{15}}{90}\mu_0 + \frac{9 + i\sqrt{15}}{90}\mu_1, \quad \text{and} \quad s_1 = \frac{9 + i\sqrt{15}}{90}\mu_0 + \frac{9 - i\sqrt{15}}{90}\mu_1$$

are a pair of Brioschi roots. Finding the other three roots now reduces to solving a cubic equation, a process that can be carried out by radicals or by further iteration as described in Section 7.3.

Recall that Chapter 5 uses radicals to reduce the general quintic to Brioschi form. The summary Figure 5.9.1 shows in detail how solving the Brioschi quintic translates back to solving the general case. We have learned that radicals alone do not suffice to solve the quintic in its general form; classically, special forms were solved with transcendental functions. In particular, Klein solved the Brioschi form a century ago with the inverse to his icosahedral invariant, a function powerful enough to combine with radicals to solve the quintic but not equations of higher degree. Now, in concert with Klein's icosahedral geometry, modern dynamical theory shows that iteration and radicals are also just strong enough to solve the quintic.

Exercises

7.10.1. Let $a \in A$ be any icosahedral face-center. Show that each coset $\widetilde{\Gamma}_T\gamma \in \widetilde{\Gamma}_T\backslash\Gamma_I$ takes a into one of the $\widetilde{\Gamma}_T$-orbits $\mathcal{O}$, $\mathcal{O}'$, $\mathcal{P}$. Show that one coset $\widetilde{\Gamma}_T\gamma_0$ takes a into $\mathcal{O}$, another coset $\widetilde{\Gamma}_T\gamma_1$ takes a into $\mathcal{O}'$, and the other three cosets take a into $\mathcal{P}$. Show that the tetrahedral subgoups containing stab(a) are explicitly $\Gamma_{T,0} = \gamma_0^{-1}\widetilde{\Gamma}_T\gamma_0$ and $\Gamma_{T,1} = \gamma_1^{-1}\widetilde{\Gamma}_T\gamma_1$.

7.10.2. Show that $\mu_{\widehat{W}}$ lies in the subring $\mathbf{C}(\widehat{W})[T]$ of $\mathbf{C}(Z)[T]$ by showing that for any $g \in \Gamma_I$, replacing Z by gZ in (7.10.1) leaves the sum unaffected.

7.10.3. (a) Confirm the values of $\mu_{\widehat{W}}(\phi_Z^{-1}(a))$ and $\mu_{\widehat{W}}(\phi_Z^{-1}(-a))$. (The first value follows from Exercise 7.10.1 and properties of $\tilde{r}$. For the second value, note that $-\mathcal{O} = \mathcal{O}'$ and $-\mathcal{P} = \mathcal{P}$.)

(b) Confirm the values of s_0 and s_1.

7.10.4. Explain why k takes the form claimed in the section. Write the system of linear equations defining the coefficients b_{ij} of k and either solve them or at least verify k.

7.10.5. Implement the Brioschi algorithm and check it, finding various roots of various Brioschi quintics.

11. Onward

For the historical context of the first six chapters of this book, see Klein's *Development of mathematics in the 19th century* [Kl 2]. Meanwhile, mathematical research continues on these topics. For example, Dummit [Du] shows how to solve a quintic by radicals when its Galois group is solvable, Buhler-Reichstein [Bu-Re] generalizes Kronecker's Theorem, Crass [Cr] discusses solving the sextic by iteration of more than one variable, and Beukers [Be] addresses the diophantine equation $x^5 + y^3 = z^2$ using the icosahedral invariants.

Nonicosahedral approaches to the quintic abound, of course. See Sturmfels [Stu] for recent results on solutions by hypergeometric series, or more generally consult the extended bibliography accompanying the poster "Solving the Quintic with Mathematica" [Wo].

Finally, Klein's original masterpiece [Kl], especially its second part, contains lovely material beyond this book that it inspired.

Index

List of symbols

Bibliography

[Ah] Lars V. Ahlfors. *Complex Analysis.* McGraw–Hill, third edition, 1979.

[Be] Frits Beukers. The diophantine equation $ax^p + by^q = cz^r$. University Utrecht Dept. Math. Preprint nr. 912 April 1995.

[Bu-Re] J. Buhler and Z. Reichstein. On the essential dimension of a finite group. Comp. Math., in press.

[Co-Li-O'S] David Cox, John Little, and Donal O'Shea. *Ideals, Varieties, and Algorithms.* Springer-Verlag, 1992.

[Co 1] H. S. M. Coxeter. *Regular Polytopes.* Methuen and Company, 1948.

[Co 2] H. S. M. Coxeter. *Regular Complex Polytopes.* Cambridge University Press, second edition, 1991.

[Cr] Scott Crass. *Solving the Sextic by Iteration: A Complex Dynamical Approach.* PhD thesis, University of California, San Diego, 1996.

[Di] L. E. Dickson. *Modern Algebraic Theories.* Sanborn, 1926.

[Do-Mc] Peter Doyle and Curt McMullen. Solving the quintic by iteration. *Acta Math.*, 163:151–180, 1989.

[Du] D. S. Dummit. Solving solvable quintics. *Math. Comp.*, 57, 1991.

[Ha] Robin Hartshorne. *Algebraic Geometry.* Springer-Verlag, 1977.

[He] I. N. Herstein. *Topics in Algebra.* Xerox, second edition, 1975.

[Hi-Co] D. Hilbert and S. Cohn-Vossen. *Geometry and the Imagination.* Chelsea, 1952.

[Ja I] Nathan Jacobson. *Basic Algebra I.* Freeman, second edition, 1985.

[Ja II] Nathan Jacobson. *Basic Algebra II.* Freeman, 1980.

[Jo-Si] Gareth A. Jones and David Singerman. *Complex Functions, an algebraic and geometric viewpoint.* Cambridge University Press, 1987.

[Kl] Felix Klein. *Lectures on the Icosahedron and the Solution of Equations of the Fifth Degree.* Chelsea, 1956. Translated by George Gavin Morrice, M. A., M. D.

[Kl 2] Felix Klein. *Development of mathematics in the 19th century.* Math. Sci. Press, 1979. Translated by M. Ackerman.

[La] Serge Lang. *Algebra.* Addison-Wesley, third edition, 1993.

[Ly-Ul] R. C. Lyndon and J. L. Ullman. Groups of elliptic linear fractional transformations. *Proc. Am. Math. Soc.*, 18:1119–1124, 1967.

[Mc] Curt McMullen. Families of rational maps and iterative root-finding algorithms. *Ann. Math.*, 125:467–493, 1987.

[Ma-Ho] Jerrold E. Marsden and Michael J. Hoffman. *Basic Complex Analysis.* Freeman, second edition, 1987.

[Mu] David Mumford. *Algebraic Geometry I: Complex Projective Varieties.* Springer-Verlag, 1976.

[Oh] Jack Ohm. On subfields of rational function fields. *Arch. Math.*, 42:136–138, 1984.

[Re-Vo] Z. Reichstein and N. Vonessen. An embedding property of universal division algebras. *J. Algebra*, 177:451–462, 1995.

[Reid] Miles Reid. *Undergraduate Algebraic Geometry.* Cambridge University Press, 1988.

[Ro] P. Roquette. Isomorphisms of generic splitting fields of simple algebras. *J. Reine Angew. Math.*, 214–215:207–226, 1964.

[Ru] Walter Rudin. *Principles of Mathematical Analysis.* McGraw-Hill, third edition, 1976.

[Sp] Michael Spivak. *Calculus on Manifolds.* Benjamin, 1965.

[Ste] Ian Stewart. *Galois Theory.* Chapman and Hall, second edition, 1989.

[Stu] Bernd Sturmfels. Solving algebraic equations in terms of a $\mathcal{A}$-hypergeometric series. Manuscript.

[Wo] Solving the quintic with mathematica. Wolfram Research, 1994.

[Za-Sa I] Oscar Zariski and Pierre Samuel. *Commutative Algebra*, volume I. Van Nostrand, 1958.